The series Lecture Notes in Computer Science (LNCS), including its subseries Lecture Notes in Artificial Intelligence (LNAI) and Lecture Notes in Bioinformatics (LNBI), has established itself as a medium for the publication of new developments in computer science and information technology research, teaching, and education.

LNCS enjoys close cooperation with the computer science R & D community, the series counts many renowned academics among its volume editors and paper authors, and collaborates with prestigious societies. Its mission is to serve this international community by providing an invaluable service, mainly focused on the publication of conference and workshop proceedings and postproceedings. LNCS commenced publication in 1973.

Alessandro Aldini
Editor

Software Engineering and Formal Methods

SEFM 2023 Collocated Workshops

CIFMA 2023 and OpenCERT 2023
Eindhoven, The Netherlands,
November 6–10, 2023
Revised Selected Papers

 Springer

Editor
Alessandro Aldini 🆔
University of Urbino Carlo Bo
Urbino, Italy

ISSN 0302-9743   ISSN 1611-3349 (electronic)
Lecture Notes in Computer Science
ISBN 978-3-031-66020-7   ISBN 978-3-031-66021-4 (eBook)
https://doi.org/10.1007/978-3-031-66021-4

Preface

The 21th International Conference on Software Engineering and Formal Methods (SEFM 2023) was organized in Eindhoven on November 6–10, 2023, by the Eindhoven University of Technology. This volume collects the proceedings of two collocated SEFM workshops:

- **CIFMA 2023**: 5th International Workshop on Cognition: Interdisciplinary Foundations, Models and Applications.
- **OpenCERT 2023**: 11th International Workshop on Open Community Approaches to Education, Research and Technology.

A single-blind peer-review process ensured the quality of the papers selected for this volume from the 16 submissions. This process was guided by the workshop organizers and managed by the Program Committees members in a rigorous way, by respecting conflict of interest situations, and with the aim of providing useful reviews to all the authors.

The workshops were held on November 6 and 7, and the discussions, always vibrant and characterized by interdisciplinary exchanges, benefited also from two keynote speeches.

We would like to thank the members of the program committees, the keynote speakers, and the authors for their effort in contributing to a rich and interesting program. We also thank the SEFM chairs for taking care of the logistics and registration process of the collocated workshops.

Urbino, Italy   Alessandro Aldini
June 2024

Contents

OpenCERT 2023—11th International Workshop on Open Community Approaches to Education, Research and Technology

CIFMA 2023—5th International Workshop on Cognition: Interdisciplinary Foundations, Models and Applications

CIFMA 2023 Organizers' Message

Cognition encompasses many aspects of intellectual functions and processes. Although it originated from the field of psychology, it goes beyond the individual human mind and behaviour, and involves and affects the interaction with the environment in which humans act. The increasing complexity of the environment with which humans interact is no longer restricted to their natural living environment and the other humans populating it, but includes a large technological support consisting of physical and computational systems, virtual worlds and robots.

This workshop investigates the study of cognition in a large number of disciplines and from different perspectives, notably in the fields of linguistics, neuroscience, psychiatry, psychology, education, philosophy, anthropology, linguistics, biology, systemics, logic and computer science. The main objective of CIFMA is bringing together practitioners and researchers from academia, industry and research institutions who are interested in the foundations and applications of cognition from the perspective of their areas of expertise. In particular, contributions to the workshop cover the areas of education, research and technology, either in general or with a focus on formal methods.

We received a total of 12 submissions about topics including logics, cognition and AI, philosophy of cognition, cognitive computing and models; 9 of these papers were accepted for presentation. The program was completed by an invited talk entitled "Deliberative Consensus", given by Davide Grossi of the Bernoulli Institute for Mathematics, Computer Science and Artificial Intelligence (University of Groningen).

We are grateful to the Program Committee for their dedication to the critical tasks of reviewing the submissions. We are also grateful to members of the Organizing Committee of SEFM for making the necessary arrangements. Finally, we thank the keynote speaker and the authors for their efforts in writing their papers and for the excellent presentations.

Alessandro Aldini
June 2024

CIFMA 2023 Organization

Program Committee Chair

Alessandro Aldini University of Urbino Carlo Bo, Italy

Publicity Chair

Pierluigi Graziani University of Urbino Carlo Bo, Italy

Program Committee

Samuel Alexander	The U.S. Securities and Exchange Commission, USA
Oana Andrei	University of Glasgow, UK
Francesco Bianchini	University of Bologna, Italy
José Creissac Campos	University of Minho, Portugal
Ana Cavalcanti	University of York, UK
Antonio Cerone	Nazarbayev University, Kazakhstan
Gustavo Cevolani	IMT School for Advanced Studies Lucca, Italy
Luisa Damiano	University of Messina, Italy
Edoardo Datteri	University of Milano-Bicocca, Italy
Yannis Haralambous	IMT Atlantique, France
Bipin Indurkhya	Jagiellonian University, Poland
Reinhard Kahle	NOVA University Lisbon, Portugal
Ulrich Kohlenbach	Technische Universität Darmstadt, Germany
Antonio Lieto	University of Turin, Italy
Mieke Massink	CNR-ISTI, Italy
Paolo Milazzo	University of Pisa, Italy
Graham Pluck	Chulalongkorn University, Thailand
Giuseppe Primiero	University of Milan, Italy
Pedro Quaresma	University of Coimbra, Portugal
Giuseppe Sergioli	University of Cagliari, Italy
Mirko Tagliaferri	University of Urbino, Italy
Marieke van Vugt	University of Groningen, The Netherlands

Additional Reviewers

Giovanna Broccia
Gabriele Ferretti

CL-XAI: Toward Enriched Cognitive Learning with Explainable Artificial Intelligence

Muhammad Suffian[1]([envelope]) [iD], Ulrike Kuhl[2] [iD], Jose Maria Alonso-Moral[3] [iD], and Alessandro Bogliolo[1] [iD]

[1] Department of Pure and Applied Sciences, University of Urbino, Urbino, Italy
m.suffian@campus.uniurb.it, alessandro.bogliolo@uniurb.it
[2] Research Institute for Cognition and Robotics, Bielefeld University, Bielefeld, Germany
ukuhl@techfak.uni-bielefeld.de
[3] Centro Singular de Investigación en Tecnoloxías Intelixentes (CiTIUS), Universidade de Santiago de Compostela, 15782 Santiago de Compostela, Spain
josemaria.alonso.moral@usc.es

Abstract. Artificial Intelligence (AI) is transforming education by providing personalized learning paths that cater to individual needs. Explainable AI (XAI) plays a vital role in assisting learners with limited prior knowledge and skills, offering transparency by elucidating how AI systems reach conclusions. In a co-learning environment, where learners engage collaboratively, the role of XAI becomes even more significant. This transparency not only fosters trust in AI-driven learning but also empowers co-learners to actively participate in and enhance their cognitive learning journey. Providing explanations for novel concepts is recognised as a fundamental aid in the learning process, particularly when addressing challenges stemming from knowledge deficiencies and skill application. Addressing these difficulties involves timely explanations and guidance throughout the learning process, prompting the interest of AI experts in developing explainer models. In this paper, we introduce an intelligent system (CL-XAI) for cognitive learning supported by XAI, focusing on two key research objectives: (i) exploring how human learners comprehend the internal mechanisms of AI models using XAI tools; and (ii) evaluating the effectiveness of such tools through human feedback. The use of CL-XAI is illustrated with a game-inspired virtual use case where learners tackle combinatorial problems to enhance problem-solving skills and deepened their understanding of complex concepts. The analysis of a pilot study of 21 participants indicates improved learning outcomes with explanations, showing a positive correlation with the complexity of tasks. Notably, over 60% of the participants expressed satisfaction, recognizing explanations as valuable aids in achieving learning targets. This paper is a pathway highlighting the potential for transformative advances in cognitive learning and co-learning.

Keywords: Cognitive learning · Explainable AI · Human-centered AI · Problem solving · Counterfactual explanations · Co-learning · Education

A. Aldini (Ed.): SEFM 2023, LNCS 14568, pp. 5–27, 2024.
https://doi.org/10.1007/978-3-031-66021-4_1

1 Introduction

In learning theory, artificial intelligence (AI), and human-computer interaction (HCI), the pursuit of problem-solving and optimal solution finding have historically been perceived through distinct lenses for machines and humans [22,44,48]. Machines, equipped with computational capabilities and cost-driven optimization, operate in a space detached from human cognition [30]. Conversely, humans rely on their unique problem-solving approaches, drawing from experiences and intuition [29]. This separation contradicts the principles of co-learning and effective HCI. In our contemporary era, where AI systems often outperform humans in various domains, bridging this gap is imperative to create a more enriching learning environment [9].

The underexplored domain of human-machine co-learning, aiming to foster mutual improvement and progress, must be addressed. Humans need an opportunity to plunge into the intricate inner workings of AI models, actively participating in co-learning to solve complex problems. In turn, AI systems should reap the benefits of human wisdom, leveraging human input and feedback to alleviate the computational burdens that once restrained their problem-solving prowess. As AI systems play an increasingly pivotal role in making high-stakes recommendations and decisions across various domains, the demand for eXplainable AI (XAI) to elucidate the rationale behind these systems grows [1,2,39]. Some scholars emphasize the collaborative aspect of explanations in XAI research [28,37,38]. There are some publications that conceptualize explanation as a two-way process, with users providing input and the XAI program providing explanatory information [12,36,37]. However, these works lack the focus on cognitive learning and its evaluation. Enhancing cognitive learning through user interaction with explanatory information is a promising avenue.

This paper introduces the so-called Cognitive Learning with XAI (CL-XAI) system. We explore the potential of co-learning with counterfactual explanations (CEs), where humans and machines collaborate in problem-solving tasks. CEs enable users to grasp the "what if" aspect of AI decisions, shedding light on alternative courses of action and improving transparent communication [46]. In this context, the focus is twofold: firstly, we investigate how human learners' cognitive learning processes are influenced when they use the CL-XAI system to receive explanations. Secondly, we assess the effectiveness of the CL-XAI system by gathering feedback from humans to evaluate its feasibility and helpfulness in providing explanations. The co-learning experience unfolds with regular interactions between learners and an explanation tool. This tool bridges human intuition and machine logic, offering learners a lifeline in their pursuit of optimal solutions. If a human learner successfully identifies an optimal solution, then it is evidence of the acquisition of knowledge comparable to that acquired by a machine learning (ML) model [32]. For the co-learning experience, we propose a virtual game-inspired scenario where learners solve combinatorial problems to achieve improved results (see Sect. 3). The learner receives explanations at regular intervals to solve the task, and the log of the learner's choices and attempts is recorded to evaluate its mental model. Overall, we believe that the synergy between human cognition and XAI guidance holds the promise of transformative advances in cognitive learning, a step towards co-learning, ultimately bolstering education on problem-solving skills and fostering a comprehensive understanding of complex concepts.

The findings derived from a user study involving 21 participants indicate a positive relationship between learning and task performance. Specifically, there is an observed upward trend in learning as users are provided with explanations. Notably, 60% of the participants expressed satisfaction with the provided explanations, recognizing them as valuable aids in achieving their respective learning targets.

In summary, the main contributions in this paper are as follows:

1. We propose the CL-XAI system that enhances the cognitive learning of user by using CEs.
2. We conducted a pilot study to evaluate the performance of the CL-XAI in terms of task performance, satisfaction, and understanding.
3. We provide empirical evidence that the CL-XAI can generate effective explanations to aid user's cognitive learning in terms of task performance, satisfaction, and understanding.
4. We implemented the CL-XAI system as open-source software to promote Open Science, and it is made publicly available to support further investigations.

The rest of the paper is structured as follows. Section 2 provides the necessary background. Section 3 outlines the CL-XAI system, elucidating its approach to incorporating XAI into the cognitive learning process. Section 4 provides empirical proof of concept of the CL-XAI system by setting up an experiment. Section 5 illustrates the results obtained from the experiments. Section 6 discusses the different factors and limitations in the context of CL-XAI, and finally, in Sect. 7, we draw conclusions from our work, emphasising potential applications and the avenues for future research for XAI-driven cognitive learning.

2   Background

Users inherently possess mental models: personal representations of the external world, shaped by individual experience, background knowledge, and connections to related concepts [16]. When they engage in tasks, these models influence information processing, shaping expectations and decisions. Interacting with an AI system offering opaque predictions prompts users to reflect on their expectations in comparison to the machine's predictions. While this substitution may lead to revised expectations and different decisions, the lack of transparency hinders a direct comparison of users' beliefs with the AI system. Explanation-enabled situational processing allows users to comprehend the system's logic, fostering a reconciliatory process between human and machine assessments [17]. In the context of cognitive learning, providing explanations becomes crucial for users to reflect on information utilization, problem-solving, and adaptation to the AI system's logic [10,24]. Cognitive learning is a pedagogical approach that emphasizes the development of comprehensive mental models among learners [15]. Such models play a pivotal role in knowledge acquisition and problem-solving, enabling individuals to navigate complex domains effectively [42]. The collaboration between humans and machines in cognitive learning, often called co-learning, has great potential [8,23]. It has been shown to enhance problem-solving skills and deepen the understanding of complex concepts [23].

From a cognitive standpoint, receiving explanations may induce two effects: (i) altering individuals' situational information processing; and (ii) adjusting their beliefs about the feature-label relationships modeled by the AI system, reflecting their mental representation of real-world processes [4]. Hence, users might employ distinct problem-solving approaches when utilizing an explanation-adjusted mental model, even in scenarios where explanations are no longer present [4].

Prior studies have introduced theoretical frameworks for symbiotic learning systems [47]. These frameworks depict a reciprocal learning process, where the learner acquires knowledge from the system, and, conversely, the system gains insights from the learner, facilitated through reinforcement learning. Further research into explainable recommendation systems has expanded to education. For instance, Barria-Pineda et al. [3] have explored the domain of recommending resources in programming classes. Tsiakas et al. [41] have investigated using cognitive training recommendations for children at primary and secondary school.

In addition, in the field of mathematics [31] and across diverse scientific domains, including chemistry [6], educators routinely employ illustrations of step-by-step solutions for a given problem. These pedagogical tools serve a dual purpose: (i) they provide solutions; and (ii) they offer explanations rooted in an expert's mental framework, which can be comprehended from a novice's standpoint [40]. The findings in [4] demonstrate that providing feature-based explanations allows AI systems to transform users' sense-making of information and enhance their understanding of the surrounding world. More specifically, explanations alter users' situational weighting of available information and prompt adjustments in their mental models. In the previously discussed publications, the process of cognitive learning is post-hoc, which means the insights (learning analytics) generated by the system have been used for a future batch of learners (no immediate feedback). Conversely, the CL-XAI integrates an XAI tool for generating CEs, which are subsequently provided to novice learners to facilitate the construction of their mental models. This method constitutes an automated system that circumvents the necessity for an expert's mental model, as it leverages the pre-existing capabilities of the XAI system.

3   The CL-XAI System

In the context of human-centered design, it is essential to delve into the psychological aspects of explanations. Figure 1 illustrates the explanation and evaluation process for the CL-XAI system. The process is adapted from [11] and the

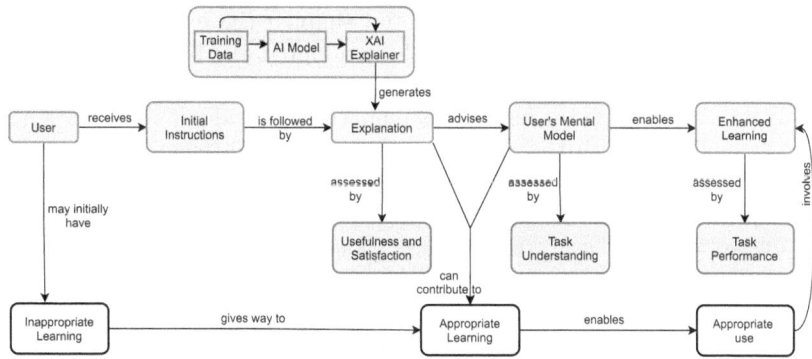

Fig. 1. Human-centered explanation and evaluation process (adapted from [11]). Blue boxes illustrate the underlying process, orange boxes illustrate objective and subjective evaluation opportunities, whereas white boxes illustrate potential learning outcomes.

modification was imperative because the initial diagram aimed to portray the entire explanation process with a focus on trust as the outcome, whereas our study specifically concentrates on evaluating cognitive learning as the ultimate result.

The CL-XAI system encompasses the following components (see Fig. 1):

1. The abstract other-wordly game scenario [20] is utilized to design the combinatorial task for the learners as a use case ("User" and "initial instructions" in Fig. 1) details in Sect. 3.1.
2. A synthetic data set is used, originally generated and statistically tested in a game-based study [20], and an ML model is trained and tested to predict the user inputs ("Training Data" and "AI Model" in Fig. 1). Further details are given in Sect. 3.2.
3. An explainer method called Diverse Counterfactual Explanations (DiCE) [27] is utilized to generate the explanations and build the user's mental model, also it assists learners in solving a given combinatorial problem (pay attention to "XAI Explainer", "Explanation", and "User's Mental Model" in Fig. 1). Further details are given in Sect. 3.3.
4. The objective and subjective evaluation measures such as task performance, task understanding, and explanation usefulness and satisfaction are employed (see orange boxes in Fig. 1). Further details are given in Sect. 3.4.
5. The technical details of the realization of the whole process for the game-based user study are given in Sect. 3.5.

3.1 Use Case and Experimental Context

The efficacy of an explanation is influenced by its purpose and the specific audience it is intended for [26]. These two factors (i.e., purpose and audience) are pivotal in selecting a suitable use case and shaping the experimental environment. Thus, in the CL-XAI system, we have adopted an abstract context called Alien Zoo [20], a virtual space wherein the learners do not have prior knowledge and only gain knowledge through the explanations provided by the CL-XAI system. Previous work with the Alien Zoo framework [20] focused on identifying factors of CEs that help or hinder interpretability, e.g., in terms of their directionality [19] or proposed plausibility constraints [18]. The current work extends this framework.

The learner's task is to nurture an Alien called 'Shub' in the Alien Zoo by feeding various combinations of plants. In our case, in contrast to the original implementation [20], the well-being of one individual Shub is directly influenced by the choices learners make in selecting plant combinations. The leaves of plants are attached with a specific time cost to find them. The learner is exposed to a combinatorial problem in which a random number of leaves of each plant are given, which are not a healthy diet for the Shub. The learner has to solve this problem by making different combinations of the plants to constitute a nourishing diet adhering to their time cost. The learner-selected plants will be fed to an AI system, which will predict whether this combination can improve Shub's health. The choices made when feeding Shub have immediate outcomes, resulting in poor or better health. The XAI tool (see Sect. 3.3) guides the learner in making optimal decisions. At regular intervals, learners are provided with explanations alongside their previous selections. These explanations highlight a choice that could have yielded a more favourable outcome. Also, these explanations are, by default, enriched with hints about the underlying data distribution associated to the AI prediction model and about how Shub's health improves, ultimately assisting learners in making informed decisions and enhancing their mental model. How we set the fictional setting in the game can be observed from the following illustrative set of instructions which are provided to the participants:

"You are the new guardian for a fictional Alien species called Shub. Different groups of Shubs live on different planets. On all planets, Shubs eat different types of leaves for their growth. But beware: each group has adapted to their own unique diet suited to their home planet. As you are new to the job, you do not know yet what works best on each planet. However, you are assisted by the [AlienNutriSolver] (XAI tool), an advanced, intergalactic dietary analysis tool. This tool uses cutting-edge algorithms to process and analyze vast arrays of data, simulating the current environment and dietary needs of Alien species across planets. To work with the AlienNutriSolver, for each planet, you have to set beforehand preliminary ranges of leaves. These ranges are taken as global constraints in the search for optimal solutions by the AlienNutriSolver. The tool will conduct a thorough exploration of potential combinations within those limits, to suggest a healthier diet for the current planet. Your task is to travel to the different planets, experiment with leaves and the AlienNutriSolver, and find the healthiest diet for the Shub".

In the provided scenario, we evaluate the performance of actual users within a task that is set in an abstract context. A notable benefit arises from the abstract nature of the task, as it eliminates any potential interference from users' prior knowledge. In the context of feeding Shubs, every user is essentially a novice, ensuring the absence of misconceptions or pre-existing beliefs.

3.2   Training Data and Models

The underlying data were generated following this pattern: The growth rate (fitness) exhibits a linear scaling between values 3–5 for plant 2, but only if plant 4 has a value ≥ 3 and plant 5 is not smaller than 3. In the initial implementation by [20], the dataset was originally created for a regression task targeting growth rate. However, we have transformed it into a classification dataset by assigning labels 1 to data points with a growth rate greater than 1.1 and 0 otherwise (the original dataset contained a growth rate 0–1.9). Consequently, in our context, the outcome label pertains to fitness (whether 1 or 0 based on the input). The synthetic dataset encompasses all feasible plant combinations, from which we select 100,000 data points with balanced classes. In order to forecast the outcome label and, consequently, the new fitness level based on user input in each trial, we employ a logistic regression (LR) model. The LR model has demonstrated satisfactory accuracy when applied to the synthetic dataset (*accuracy:0.97, recall:0.91, and F1-score:0.88*), and it enables efficient computation of CEs. Throughout the implementation, to ensure consistent model outputs for all users across the experiment, we utilize the same LR model constructed at the outset.

3.3   DiCE

The field of XAI encompasses various technical approaches aimed at enhancing the transparency, interpretability and explainability of AI systems [2]. Human-in-the-loop interactions and natural language explanations contribute to the holistic understanding of AI operations, catering to various interpretation needs and promoting trust in AI decision-making [39]. CEs provide insights into how alterations in input data affect the model's outputs. In this context, DiCE herein referred to as the "XAI tool" [27] provides CEs[1] to the learner to devise different strategies to solve the task at hand. This XAI tool is not self-driven; rather, it exploits learner input to generate customized explanations.

3.4   Evaluation Measures

In evaluating XAI systems, examining specific cognitive states or processes, herein referred to as "cognitive metrics", is a central concern [14]. These metrics

---

[1] In the rest of the paper, we use explanations and counterfactual explanations (CEs) interchangeably.

are instrumental in assessing whether learners have achieved a pragmatic understanding of the AI system, particularly in light of the explanations furnished by the XAI tool. Drawing from established approaches in cognitive science and psychology [14], Hoffman et al. [13] proposed a conceptual model elucidating the explanation processes of an XAI system and how a learner's pragmatic comprehension of these explanations can be assessed across distinct functional stages. Hoffman's model delineates three pivotal functional stages within the operation of a XAI system: (i) *explanation generation*; (ii) *learner's mental model generation*; and (iii) *learner's enhanced performance resulting from the assimilated mental model*. Consequently, the evaluation of XAI systems through cognitive metrics can be systematically framed within these three stages, offering a comprehensive approach to gauge the effectiveness of these systems in facilitating user understanding and performance improvement.

In accordance with Hoffman et al. [13], during the explanation generation stage, an assessment of the learner's practical comprehension of the AI system can be made by examining the learner's cognitive processes, which gauge the quality of the explanation (referred to as "Explanation Goodness") and the degree of satisfaction with it (referred to as "User Satisfaction"). Between the explanation generation stage and the subsequent stage of constructing the learner's mental model, which is influenced by the explanations received, learners gradually update their mental models, with several psychological factors potentially influencing the model-building process. Evaluating the extent of a learner's understanding and satisfaction poses a formidable challenge. To address this challenge, we have extended the Alien Zoo [20], where the learners are now assigned problem-solving tasks, and their performance in solving these tasks serves as a metric to determine the level of understanding of the provided explanations. This, in turn, corroborates the effectiveness of the XAI tool in generating high-quality explanations. User Satisfaction can be assessed using subjective measures through the administration of a questionnaire, with, for example the *Explanation Satisfaction Scale* introduced by Hoffman et al. [13] and later refined by van der Waa et al. [45]. This scale provides a reliable and psychometrically robust means of gauging user satisfaction with a system's explanations.

User understanding, in the context of XAI, pertains to the development of a learner's "mental model" of a system's inner workings [14]. The concept of a "mental model" draws from psychological theories, denoting an individual's internal representation of the people, objects, and environments with which she interacts [33,35]. In the realm of XAI, for our case, an ideal outcome is for the learner's mental model to reflect the system explained via the XAI tool. Explanations play a pivotal role in facilitating the construction of precise mental models, which can be categorized as follows: (i) global understanding, signifying a general comprehension of a system's functioning; (ii) local understanding, signifying insight into a specific decision made by the system; and (iii) functional understanding, representing a grasp of the system's capabilities and intended uses [13,14]. In this work, our goal is to capture the local understanding of the learner's mental model.

3.5   Technical Details

The realization of the Alien Zoo entails a rigorous segregation between the front end, responsible for crafting the game interface that participants interact with, and the back end, delivering the predictions made by the AI system along with the explanations provided by the XAI tool. The web interface utilizes Phaser3[2], an HTML5 game framework driven by JavaScript. The system's back end, on the other hand, is founded on Python3, leveraging the sklearn[3] package for supporting ML algorithms. The underlying ML model is trained using synthetic plant data [20], and it predicts the fitness of the Shub, thereby influencing the game's dynamics. The learner input reaches this model via the front end, allowing an analysis of the potential for yielding positive outcomes and consequently enhancing the Shub's fitness. This Python-based framework is adept at generating CEs with DiCE to ensure adaptability and accommodate various ML algorithms. We implemented the CL-XAI system as open-source software to promote Open Science, and further details to reproduce it can be found at the GitHub repository using the following link (https://github.com/msnizami/Cognitive-Learning-with-XAI).

4   Empirical Proof of Concept

In the upcoming sections, we undertake an empirical examination of the effectiveness and practicality of the CL-XAI system. Namely, we conduct a user study that investigates the influence of offering CEs on cognitive learning within the iterative learning task established by the CL-XAI system.

4.1   Hypotheses

The central inquiry in this proof-of-concept study is whether users profit from receiving CEs during the process of learning and identifying relationships within an unfamiliar dataset while engaging with the CL-XAI system.

   To address this inquiry, we employ an interactive learning task wherein users iteratively choose input values for an ML model. Throughout the experiment, users receive explanations and provide suggestions to the system. The presentation illustrates how adjustments to their previous choices might have resulted in improved outcomes. This approach, involving repeated interactions and user actions, enables us to objectively assess the system's understanding through task performance and subjectively assess the usefulness and satisfaction of provided explanations across a series of tasks. Consequently, we address the following two hypotheses:

1. **H-1: Effective Cognitive Learning**. We anticipate that strong user comprehension, facilitated by the provided explanations, will result in effective

---

[2] https://phaser.io/.
[3] https://scikit-learn.org/stable/.

cognitive learning. Users will also gain the ability to explicitly recognize relevant and irrelevant input features, indicating proficient capability to discern critical data factors. Evaluation of cognitive learning involves metrics derived from task performance, including objective scores obtained during the game, as well as self-reported responses in the post-game survey.

2. **H-2: User Satisfaction and Understanding**. We anticipate that users can gain a comprehensive understanding of intricate data relationships thanks to the provision of explanations. Our goal is to assist users in gaining a profound understanding of the system's complexities by offering CEs. The post-game survey is employed to gather feedback from learners regarding their satisfaction with the explanations. Learners' evaluations of the explanations' utility and overall satisfaction contribute valuable insights into user contentment.

By examining these hypotheses, we aim to uncover how explanation quality influences cognitive learning and co-learning mechanisms with the CL-XAI system.

4.2   Experimental Setup

The experiment is composed of two phases: a game and a survey. Upon accepting the invitation, participants are redirected to a web server hosting the study.

Initially, users are informed about the study's purpose, procedure, expected duration, their right to withdraw, confidentiality, and the contact details of the primary investigator. If a user chooses not to participate, she has the option to close the window. Alternatively, users express their agreement by pressing a button. A subsequent page provides detailed information about the game, including images of the Shub to be fed and the various plants available for feeding.

Written instructions specify that the fitness of the Shub can be influenced by choosing healthy or unhealthy combinations of leaves per plant. The maximum number of leaves per plant is limited to six, and users have the freedom to select any preferred combination of plants. Following this, suggestions guide users to maximize the fitness level of the Shub to enhance learning for the task. Additionally, users receive feedback on which choices could have led to better results.

Clicking the "Start" button at the end of the page signals that the user is prepared to begin the game phase.

Participants. In mid January 2024, the study was carried out by distributing a public link to the web server hosting the game study among the colleagues in University of Urbino, University of Santiago De Compostela, and Bielefeld University. A total of 31 participants took part, providing informed electronic consent through a click-wrap agreement before engaging in the study.

Game Phase. The user interface (UI) is meticulously designed to offer learners an intuitive and engaging experience throughout their journey of nurturing a Shub and exploring the intricate relationships between plant combinations and

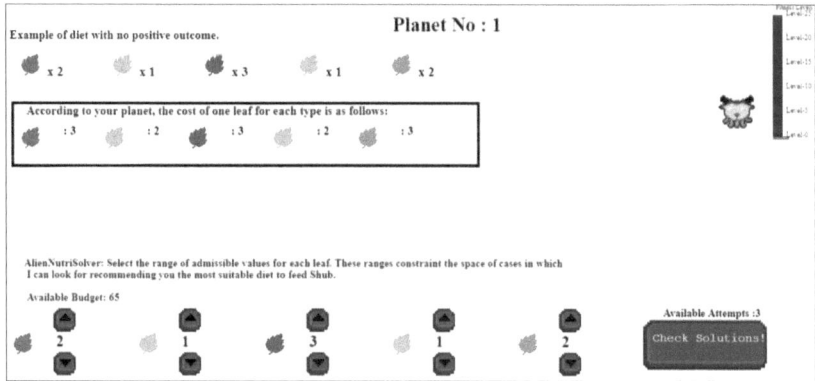

Fig. 2. The learner's task is to go for those combinations of plants by selecting from drop-down menus that can be searched in the available budget. After each search/selection, the learner feeds the plants to Shub and waits for the outcome.

growth outcomes. The UI comprises various screens, each carefully crafted to provide essential information and interactivity. In the right-hand side of Fig. 2 (configuration screen), an avatar represents a Shub and dynamically reflects its fitness level on a vertical bar as the chosen plant combination impacts the Shub's health. The bar limits indicate the optimal and unsatisfactory fitness levels. On the left, the five different plants are displayed on top, which constitute the diet, and a test input is shown just below, which needs the learner's attention to customize it by selecting any combination from the drop-down menu's given for each plant at the bottom of the screen. The user can also see the available attempts and available budget to improve the health and the cost in terms of time required to invest in one leaf of each plant. It is worth noting that the costs of leaves vary on different planets. When the user increases the number of leaves, the cost of those leaves is deducted from the budget, and the updated budget is then shown. At the start of the game, the user was instructed to be careful about the costs and budget while selecting the leaves. Additionally, the page features a "Help" button that becomes available when the user's provided inputs do not result in an improved diet. Upon pressing the Help button, the system provides suggestions to guide the user towards better diet plans. Participants are initially assigned a Shub with a fitness level of 0. To submit their choices, participants click a "Check Solution!" button in the bottom right corner of the screen, which, upon pressing, changes to "Loading." The illustration of different game screens is provided in Supplementary Material A.1. The underlying ML model predicts the new fitness based on the user's input. Subsequently, our implementation computes a CE and presents it to the user for reviewing and feeding it to Shub. Upon pressing the "Feeding Time!" button, a brief progress scene is displayed, during which the underlying ML model updates the Shub's fitness score. The user is then directed to the next planet (configuration screen), visualizing the impact of the current choice through written information and an animated Shub. Participants play the game for three planets.

Survey Phase. Participants respond to a series of questions designed to gauge various aspects. Initially, survey items aim to evaluate users' explicit knowledge regarding the relevance of plants for task success (covered in items 1 and 2 of the post-game survey). Subsequently, participants provide subjective judgments on the usefulness and satisfaction derived from the experience (covered in items 5 and 6 of the post-game survey). The rest of six self-reporting measures are employed to assess potential confounding factors. These measures focus on determining whether users comprehend the feedback provided and how they perceive the timing and effectiveness of the feedback. The last six items of the survey phase focus on gathering demographic information, specifically the participant's gender and age, education level and background, region, and English language proficiency. Upon reaching the final page of the study, participants receive gratitude for their participation. Furthermore, participants have the option to follow a link that provides comprehensive debriefing information. Further details on the survey items are provided in the Supplementary Material A.2.

5   Results

This section explores the suitability of the proposed CL-XAI system in examining the impact of offering automatically generated CEs to users tasked with learning about unknown relationships in a dataset. We collected data from 31 participants, among whom 21 played the game and responded to the post-game survey completely. Therefore, we only included data from participants who both fully engaged in the game and completed the post-game survey (see Table 1).

Table 1. Demographic details.

Gender		Age	
Male	11	18–24 year	1
Female	6	25–34 year	11
Not listed	1	35–44 year	4
Prefer not to say	3	Prefer not to say	5

5.1   H-1: Effective Cognitive Learning

The hypothesis H-1 asserts that strong user comprehension, facilitated by the provided explanations, results in effective cognitive learning. The evaluation of cognitive learning involves metrics derived from task performance, including objective scores obtained during the game, as well as self-reported responses

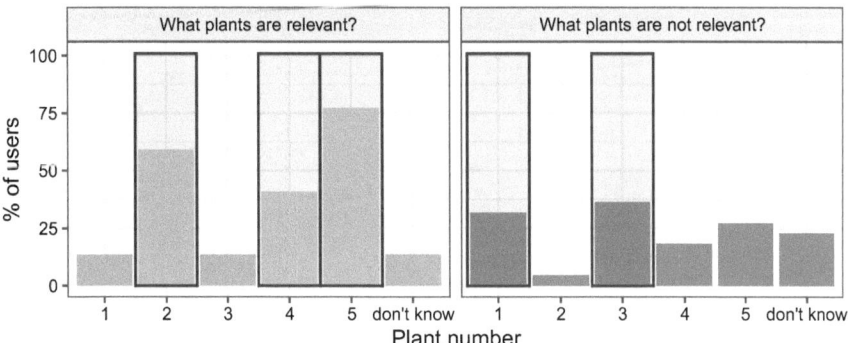

Fig. 3. H-1: the judged relevance of plants by the users, the blue rectangular boxes represent the ground truth plants.

in the post-game survey. We analyzed the objective scores from the user gameplay (i.e., mean fitness score and mean number of attempts per planet) and self-reported responses in the post-game survey to see whether the users were able to explicitly recognize relevant and irrelevant input features. For hypothesis H-1, we performed two analyses and present them in the following as *Judged relevance of features* and *Task performance*.

Judged relevance of plants. The plots in Fig. 3 illustrates users' perceived relevance of various plants (features). In the left plot, blue rectangular boxes delineate the ground truth of plants 2, 4, and 5 deemed relevant for the Shub's fitness. Conversely, in the right plot, plants 1 and 3 were deemed irrelevant, prompting the delineation of blue rectangular boxes around them. Analysis reveals that over 75% of users correctly identified plant 5, with 60% recognizing plant 2, and 40% accurately pinpointing plant 4. Conversely, in the right plot, over 30% and 35% of users correctly identified the irrelevant plants 1 and 3, respectively. These findings suggest that the provided explanations were effective for users' cognitive learning, thereby fostering a good understanding of different factors crucial for optimal fitness attainment.

Task performance. In the game, there were three tasks, each to be performed on a specific *[planet]* (i.e., task-1 on planet-1, task-2 on planet-2, and task-3 on planet-3). The users were assigned a consolidated budget for all tasks to select leaves for constituting the diet for the Shub. The difficulty level of tasks was increasing from planet-1 to planet-3. This was due to a systematic change on cost function: the costs for relevant plants increase throughout the game, forcing users to make more economical decisions in the later stages. Additionally, the reduction of the available budget due to decisions in earlier stages of the game may further complicate decision-making, punishing participants who were too generous and/or explorative earlier on. For successful completion of one

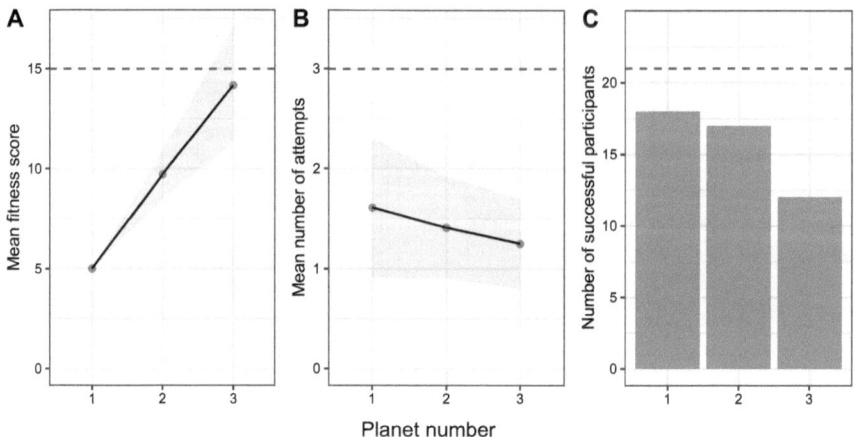

Fig. 4. H-1: Users' objective task performance over the journey of three planets. (A) Mean fitness score achieved for the subgroup of participants who were able to successfully improve the Alien's fitness (i.e., successful participants). (B) Average number of attempts needed by successful participants. (C) Visualization of participant attrition over the course of the study. Shaded areas in (A, B) denote the standard deviation. Dashed red lines illustrate maximal achievable score (A), maximal number of attempts possible (B), and number of participants for whom data was available (C).

task (i.e., one planet), the system credits the user a score of 5. Thus, after the successful completion of all three tasks it will assign the user a score of 15. Figure 4 illustrates users' objective task performance over the journey of three tasks in terms of average fitness score achieved (Fig. 4A), and mean number of attempts needed (Fig. 4B) per planet, for those participants who were able to successfully finish the tasks, respectively.

Predictably, mean fitness for successful participants reliably increases throughout all three tasks. Still, it is striking that there is quite some variation between successful users, indicated by the shaded area in Fig. 4A. This clearly illustrates that users that solve the task in later stages may not possess the full final score of 15, simply due to not achieving points in the initial phase of the game. This insight emphasizes how achieving proficiency in an iterative learning task as presented here is a gradual process. As we move from planet 1–3, the mean number of attempts needed by successful participants decreases, indicating a positive learning curve: Users that succeed in the task successively gain greater insight into the underlying data distribution, thus becoming more proficient in selecting a healthy diet right away. Finally, the number of successful participants reduces throughout the game (Fig. 4C), reflecting the increased task difficulty. The increasing cost triggers participant attrition in later stages of the game. Failure to achieve the maximum score could be ascribed to individual playing style, spending higher costs for the selection of leaves in the first task and thus reducing the available budget for upcoming tasks. In severe cases, this strategy might hinder users to select the relevant leaves in the final task to finish the task successfully in the last stage. Still, we observe that more than half

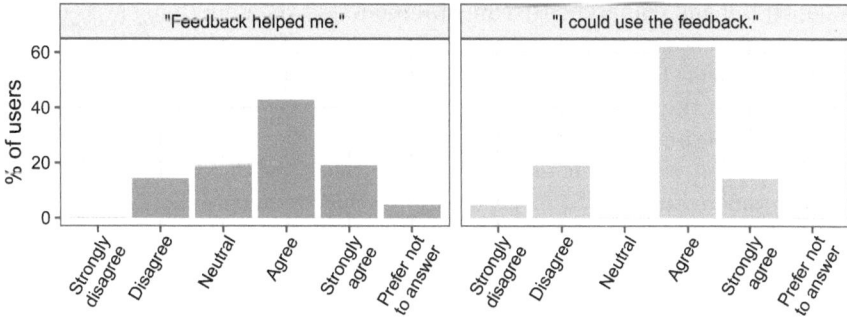

Fig. 5. H-2: Percentage of user responses to survey items assessing the helpulness (left) and usability (right) of CEs, as assessed in the post-game survey.

of the users were able to achieve the maximum score by the end of the game, supporting our hypothesis H-1, positing that the provided explanations resulted in effective cognitive learning.

5.2   H-2: User Satisfaction and Understanding

The hypothesis H-2 states that users can gain a comprehensive understanding of intricate data relationships through the provision of explanations. We analyzed the results of the post-game from all learners, irrespective of their success in the game, regarding their satisfaction with the explanations. The plots in Fig. 5 illustrate users' self-reported understanding and satisfaction. The analysis reveals that the corresponding item ("I could use the feedback.") got positive response of over 60% of users. Similarly, over 40% of users reported positive response to item ("Feedback helped me."). These results show that users can constitute a good diet for the Shub's fitness but also reflects their understanding about the underlying task difficulty. Thus, users were satisfied with the provided explanations, thereby endorsing the effective usability of explanations.

6   Discussion

In the proof-of-concept pilot study, we explored the effectiveness and feasibility of the CL-XAI system in the context of cognitive learning. Specifically, we investigated how providing CEs influences cognitive learning, gauging user performance through objective behavioral metrics and subjective self-reports. Our findings highlight the CL-XAI system's potential in studying strategies for supporting cognitive learning.

Despite the interdependence of three features in the underlying data, participants successfully increased the health of the Shub throughout the experiment. In addition, users consistently identified the relevant plants for the task over multiple attempts. However, this should not be construed as evidence that users developed full and comprehensive mental models of the underlying system, given the substantial improvement in task performance. For example, 75% of all users in the experiment identified plant 5 as a relevant feature, while only about 30% also recognized the importance of plant 4. The provision of CEs not only affects the learner's performance but also contributes to the subjective comprehension of the system. A significant proportion of learners responded positively regarding their understanding of the explanations. Survey items illustrated in Figs. 3 and 5 reflect the impact of the effectiveness of the CL-XAI system.

It is crucial to recognize and address the limitations inherent in this proof-of-concept study. While we successfully demonstrate the benefits of providing CEs as feedback in an iterative learning design tailored for an abstract domain and novice users, the generalizability of this observation to other tasks, domains, and target groups is limited [34]. CEs function as local explanations, focusing on rectifying past predictions. Consequently, it is unlikely for users to form an accurate mental model of the entire underlying system based solely on a sparse set of these specific explanations, what presents a limitation given the importance of completeness in this process [21]. Future research should explore scenarios that significantly impact the usability of CEs, especially when they cannot offer a comprehensive picture.

Moreover, we do not investigate whether the provision of CEs also enhances users' trust in the system. Trust is a crucial factor in XAI, extensively studied in the literature [7]. While the current study exclusively focuses on the learning aspect, the setup could be extended to include the evaluation of trust, for instance, by incorporating corresponding items into the survey. In cognitive psychology research, learners' performance in task-learning endeavors is subject to the influence of many variables. These encompass factors such as age [5], learning experience [44], cognitive abilities [15], cultural and socio-emotional factors [43]. Factors related to mental health [25] collectively constitute critical determinants.

7   Conclusion

In this paper, we have introduced the CL-XAI system, designed to facilitate cognitive learning with XAI. The outcomes of our research offer significant potential for enhancing the development and refinement of XAI techniques, ultimately leading to improved cognitive learning experiences.

The deliberate focus on the convergence of XAI and cognitive learning stems from the recognition that learners, especially in educational and training contexts, stand to gain substantially from AI systems that are understandable and transparent. Our work, centered on providing timely and lucid explanations for intricate concepts and problem-solving tasks (combinatorial problems), seeks to empower learners, bridge knowledge disparities, and cultivate a deeper understanding of a challenging subject matter. Furthermore, the implications of this research extend beyond the realm of education to encompass domains where human-AI collaboration is pivotal, such as healthcare diagnostics, legal decision support, and financial analysis.

In summary, we assert that the synergy between human cognition and the guidance offered by the CL-XAI system holds the potential for transformative advancements in cognitive learning. This will mark a significant stride toward co-learning, ultimately fortifying problem-solving abilities and nurturing a comprehensive grasp of complex concepts.

Acknowledgments. This research was funded by MCIN/AEI/10.13039/5011 00011033 (grants PID2021-123152OB-C21 and TED2021-130295B-C33), the Galician Ministry of Culture, Education, Professional Training, and University (grants ED431C2022/19 and ED431G2019/04). All grants were co-funded by the European Regional Development Fund (ERDF/FEDER program). Ulrike Kuhl was supported by the research training group "Dataninja" (Trustworthy AI for Seamless Problem Solving: Next Generation Intelligence Joins Robust Data Analysis) funded by the German federal state of North Rhine-Westphalia.

A   Supplementary Material

A.1   Exemplary User Learning Journey in the Game Phase

An exemplary user learning journey for planet-1 (task-1, and this is same for each planet) is illustrated in Figs. 7, 6, 8, 9, 10 and 11. As it can be seen in Fig. 7, the user begins with defining plant ranges to find the good solution (diet) for the better fitness of the Shub. Then, as it is depicted in Fig. 6 the user receives a feedback, since in this case the selection was not finding a valid solution. Then, the user is provided some suggestion after clicking on the Help button as indicated in Fig. 8. After receiving the suggestions, the user has to re-select the plant ranges for finding a good solution as indicated in Fig. 9. After this, the user feeds the found solution as a diet and a progress screen displays that diet is in feeding process (Fig. 10). Finally, the user is greeted and instructed to move to the next planet (Fig. 11).

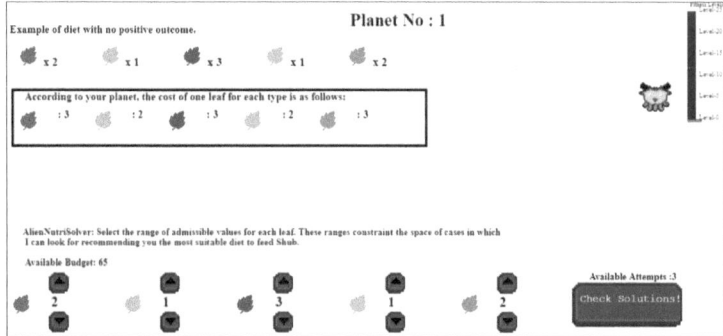

Fig. 6. Game Phase: Screen-2, the user selected ranges to constitute the diet resulted into no valid solution for better diet. The user receives feedback to look for suggestions from the Help button, and to try again.

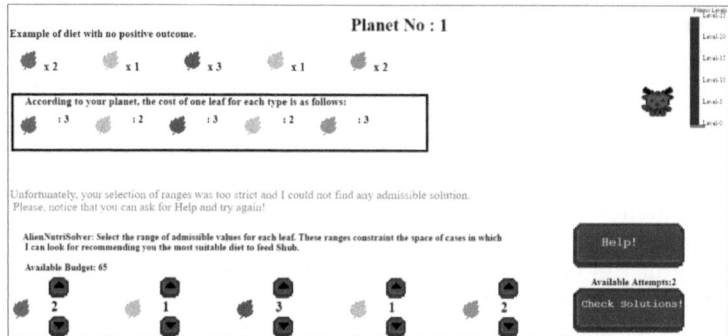

Fig. 7. Game Phase: Screen-1, an example diet predicted as not good diet for the fitness of the Shub is shown, the user can play with an alternative diet by selecting the ranges for leaves from the drop-down menu keeping in view of the cost of leaves and available budget.

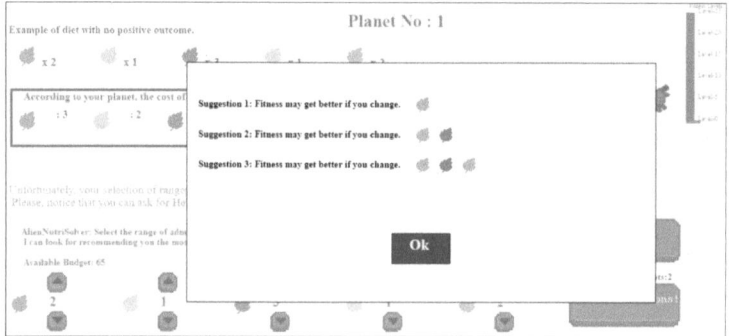

Fig. 8. Game Phase: Screen-3, the display of suggestions when the user click on Help button.

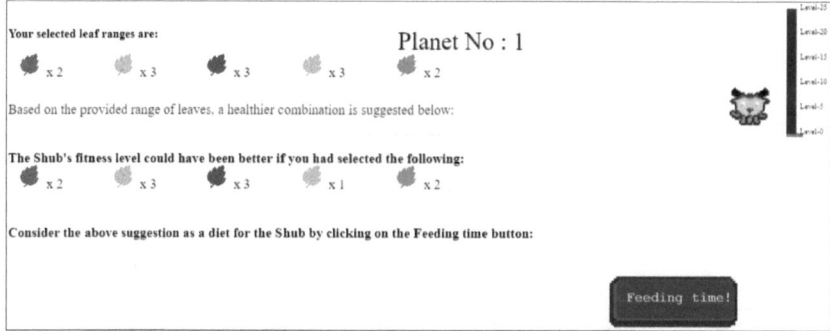

Fig. 9. Game Phase: Screen-4, the user selected ranges which resulted into a good diet and it is suggested to the user for feeding to the Shub.

Fig. 10. Game Phase: Screen-5, when the user click on the "Feeding time" button, this screen shows that feeding is in progress, it means the model is computing the new fitness level for the Shub.

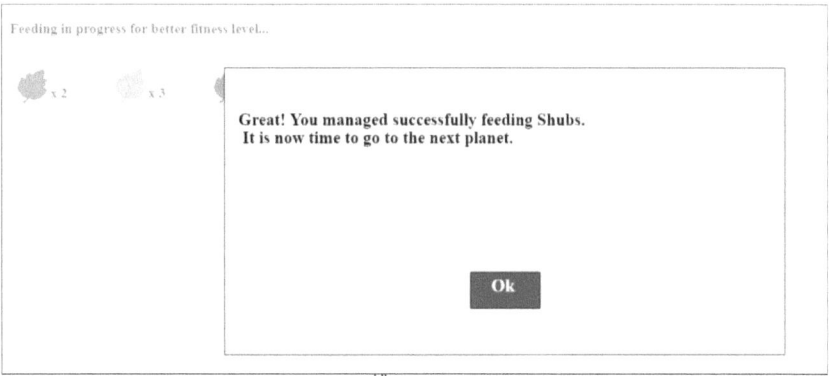

Fig. 11. Game Phase: Screen-6, on feeding good diet to the Shub, the user is greeted and instructed to move to the next planet (task).

A.2   Survey Items Table

In Table A.2, the different post-game survey items are presented with their item number, item statement, and response options. These items were used to endorse the hypotheses established in the main content of the paper. For example, items 1 and 2 were employed for hypothesis H-1 and items 5 and 6 for hypothesis H-2. The age and the gender items were only used from the demographic items in the paper for the participants of the study (Table 2).

Table 2. Table of survey items.

Item no	Item statement	Response options
1	What do you think: Which plants were relevant to increase the fitness of Shub? Please select ALL that you think were relevant	5 checkboxes, together with icons of the available plants + option "I do not know."
2	What do you think: Which plants were irrelevant to increase the fitness of Shub? Please select ALL that you think were relevant	=
3	I understood the feedback on what choice would have led to a better result.	5 point Likert-scale, checkboxes with options: Strongly disagree - disagree - neutral - agree - strongly agree + option "I prefer not to answer."
4	I needed support to understand the feedback on what choice would have led to a better result	=
5	I found that the feedback on what choice would have led to a better result helped me to increase the fitness of Shub	=
6	I was able to use the feedback on what choice would have led to a better result to increase the fitness of Shub	=
7 (catch item)	Are you still paying attention? If so, please select 'I prefer not to answer' for this question	=
8	I found inconsistencies in the feedback on what choice would have led to a better result	=
9	I think most people would learn to work with the feedback on what choice would have led to a better result very quickly	=
10	I received the feedback on what choice would have led to a better result in a timely and efficient manner	=
Age	Please indicate your age	Checkboxes with options: 18–24y, 25–34y, 35–44y, 45–54y, 55–64y, 65y and over
Gender	Which term most accurately describes your gender?	Checkboxes with options: Female, Male, Transgender female, Transgender male, Non-binary/gender non-conforming, Not listed, I prefer not to answer
Education	What is your education level?	Checkboxes with options: Secondary, upper-secondary, short-cycle tertiary, Bachelors, Masters. Doctorate, I prefer not to answer
Background	What is your education background (broader discipline)?	Checkboxes with options: Science , Technology, Engineering, Arts, Social sciences, Humanities I prefer not to answer
Region	What is your geographical region?	Checkboxes with options: Europe, United kingdom, United states, South-asia, China-Japan, Middle-east I prefer not to answer
Education	What is your English language proficiency level?	Checkboxes with options:Beginner(A1), Elementary(A2), Intermediat(B1), upper-intermediate(B2), Advanced(C1), Native, I prefer not to answer

References

1. Adadi, A., Berrada, M.: Peeking inside the black-box: a survey on explainable artificial intelligence (XAI). IEEE Access **6**, 52138–52160 (2018)
2. Ali, S., Abuhmed, T., El-Sappagh, S., Muhammad, K., Alonso-Moral, J.M., Confalonieri, R., Guidotti, R., Ser, J.D., Díaz-Rodríguez, N., Herrera, F.: Explainable artificial intelligence (XAI): what we know and what is left to attain trustworthy artificial intelligence. Inf. Fus. 101805 (2023). https://doi.org/10.1016/j.inffus.2023.101805
3. Barria-Pineda, J., Akhuseyinoglu, K., Želem-Ćelap, S., Brusilovsky, P., Milicevic, A.K., Ivanovic, M.: Explainable recommendations in a personalized programming practice system. In: International Conference on Artificial Intelligence in Education, pp. 64–76. Springer (2021)
4. Bauer, K., von Zahn, M., Hinz, O.: Expl(AI)ned: The impact of explainable artificial intelligence on users' information processing. Inf. Syst. Res. (2023)
5. Chan, C.Y.H., Chan, A.B., Lee, T.M.C., Hsiao, J.H.: Eye-movement patterns in face recognition are associated with cognitive decline in older adults. Psychon. Bull. Rev. **25**, 2200–2207 (2018)
6. Crippen, K.J., Earl, B.L.: The impact of web-based worked examples and self-explanation on performance, problem solving, and self-efficacy. Comput. Educ. **49**(3), 809–821 (2007)
7. Davis, B., Glenski, M., Sealy, W., Arendt, D.: Measure utility, gain trust: practical advice for XAI researchers. In: IEEE Workshop on Trust and Expertise in Visual Analytics (TREX). pp. 1–8. IEEE (2020)
8. Deiss, O., Biswal, S., Jin, J., Sun, H., Westover, M.B., Sun, J.: HAMLET: interpretable human and machine co-learning technique (2018). arXiv preprint arXiv:1803.09702
9. Grace, K., Salvatier, J., Dafoe, A., Zhang, B., Evans, O.: When will AI exceed human performance? Evidence from AI experts. J. Artif. Intell. Res. **62**, 729–754 (2018)
10. Gregor, S.: The nature of theory in information systems. MIS Q. 611–642 (2006)
11. Gunning, D., Vorm, E., Wang, J.Y., Turek, M.: Darpa's explainable AI (XAI) program: a retrospective. Appl. AI Lett. **2**(4) (2021).https://doi.org/10.1002/ail2.61
12. Hoffman, R.R., Miller, T., Klein, G., Mueller, S.T., Clancey, W.J.: Increasing the value of XAI for users: a psychological perspective. KI-Künstliche Intelligenz 1–11 (2023)
13. Hoffman, R.R., Mueller, S.T., Klein, G., Litman, J.: Metrics for explainable AI: challenges and prospects (2018). arXiv preprint arXiv:1812.04608
14. Hsiao, J.H., Ngai, H.H.T., Qiu, L., Yang, Y., Cao, C.C.: Roadmap of designing cognitive metrics for explainable artificial intelligence (XAI) (2021). arXiv preprint arXiv:2108.01737
15. Johnson-Laird, P.N.: Mental models and cognitive change. J. Cogn. Psychol. **25**(2), 131–138 (2013)
16. Jones, N.A., Ross, H., Lynam, T., Perez, P., Leitch, A.: Mental models: an interdisciplinary synthesis of theory and methods. Ecol. Soc. **16**(1) (2011)
17. Kao, C.H., Feng, G.W., Hur, J.K., Jarvis, H., Rutledge, R.B.: Computational models of subjective feelings in psychiatry. Neurosci. Biobehav. Rev. **145**, 105008 (2023)
18. Kuhl, U., Artelt, A., Hammer, B.: Keep your friends close and your counterfactuals closer: improved learning from closest rather than plausible counterfactual explanations in an abstract setting. In: Proceedings of the 2022 ACM Conference on

Fairness, Accountability, and Transparency, pp. 2125-2137. Association for Computing Machinery, New York, NY, USA (2022).https://doi.org/10.1145/3531146. 3534630

19. Kuhl, U., Artelt, A., Hammer, B.: For better or worse: the impact of counterfactual explanations' directionality on user behavior in xai. In: Longo, L. (ed.) Explainable Artificial Intelligence, pp. 280–300. Springer Nature Switzerland, Cham (2023)

20. Kuhl, U., Artelt, A., Hammer, B.: Let's go to the Alien Zoo: introducing an experimental framework to study usability of counterfactual explanations for machine learning. Front. Comput. Sci. **5**, 20 (2023)

21. Kulesza, T., Stumpf, S., Burnett, M., Yang, S., Kwan, I., Wong, W.K.: Too much, too little, or just right? Ways explanations impact end users' mental models. In: IEEE Symposium on Visual Languages and Human Centric Computing. pp. 3–10. IEEE (2013)

22. Langley, P.: Intelligent behavior in humans and machines. Adv. Cogn. Syst. **2**, 3–12 (2007)

23. Lieto, A., Radicioni, D.P.: From human to artificial cognition and back: new perspectives on cognitively inspired AI systems. Cogn. Syst. Res. **39**, 1–3 (2016)

24. Malle, B.F.: How the Mind Explains Behavior: folk Explanations, Meaning, and Social Interaction. MIT press (2006)

25. Marin, M.F., Lord, C., Andrews, J., Juster, R.P., Sindi, S., Arsenault-Lapierre, G., Fiocco, A.J., Lupien, S.J.: Chronic stress, cognitive functioning and mental health. Neurobiol. Learn. Mem. **96**(4), 583–595 (2011)

26. Mohseni, S., Zarei, N., Ragan, E.D.: A multidisciplinary survey and framework for design and evaluation of explainable AI systems. ACM Trans. Interact. Intell. Syst. (TiiS) **11**(3–4), 1–45 (2021)

27. Mothilal, R.K., Sharma, A., Tan, C.: Explaining machine learning classifiers through diverse counterfactual explanations. In: Proceedings of the Conference on Fairness, Accountability, and Transparency, pp. 607-617. ACM, New York, NY, USA (2020). https://doi.org/10.1145/3351095.3372850

28. Mueller, S.T., Veinott, E.S., Hoffman, R.R., Klein, G., Alam, L., Mamun, T., Clancey, W.J.: Principles of explanation in human-AI systems. In: Proceedings of the AAAI Workshop on Explainable Agency in Artificial Intelligence (AAAI-2020) (2021)

29. Newell, A., Simon, H.A.: Computer science as empirical inquiry: symbols and search. In: ACM Turing Award Lectures, pp. 1975. New York, NY, USA (2007)

30. Newell, A., Simon, H.A., et al.: Human Problem Solving, vol. 104. Prentice-hall Englewood Cliffs, NJ (1972)

31. Renkl, A.: Learning from worked-examples in mathematics: students relate procedures to principles. ZDM **49**(4), 571–584 (2017)

32. Ribeiro, M.T., Singh, S., Guestrin, C.: Model-agnostic interpretability of machine learning (2016). arXiv preprint arXiv:1606.05386

33. Richardson, G.P., Andersen, D.F., Maxwell, T.A., Stewart, T.R.: Foundations of mental model research. In: Proceedings of the International System Dynamics Conference, pp. 181–192. EF Wolstenholme (1994)

34. Sokol, K., Flach, P.: Explainability fact sheets: A framework for systematic assessment of explainable approaches. In: Proceedings of the Conference on Fairness, Accountability, and Transparency, pp. 56–67 (2020)

35. Staggers, N., Norcio, A.F.: Mental models: concepts for human-computer interaction research. Int. J. Man Mach. Stud. **38**(4), 587–605 (1993)

36. Stepin, I., Suffian, M., Catala, A., Alonso-Moral, J.M.: How to build self-explaining fuzzy systems: from interpretability to explainability [AI-eXplained]. IEEE Comput. Intell. Mag. **19**(1), 81–82 (2024). https://doi.org/10.1109/MCI.2023.3328098

37. Suffian, M., Graziani, P., Alonso, J.M., Bogliolo, A.: FCE: feedback based counterfactual explanations for explainable AI. IEEE Access **10**, 72363–72372 (2022). https://doi.org/10.1109/ACCESS.2022.3189432

38. Suffian, M., Khan, M.Y., Bogliolo, A.: Towards human cognition level-based experiment design for counterfactual explanations. In: Mohammad Ali Jinnah University International Conference on Computing (MAJICC). pp. 1–5. IEEE (2022)

39. Suffian, M., Stepin, I., Alonso-Moral, J.M., Bogliolo, A.: Investigating human-centered perspectives in explainable artificial intelligence. In: CEUR Workshop Proceedings, vol. 3518, pp. 47–66 (2023)

40. Sweller, J.: The worked example effect and human cognition. Learn. Instr. **16**, 165–169 (2006). https://doi.org/10.1016/j.learninstruc.2006.02.005

41. Tsiakas, K., Barakova, E., Khan, J.V., Markopoulos, P.: BrainHood: towards an explainable recommendation system for self-regulated cognitive training in children. In: Proceedings of the 13th ACM International Conference on Pervasive Technologies Related to Assistive Environments, pp. 1–6 (2020)

42. VanLehn, K.: Cognitive skill acquisition. Annu. Rev. Psychol. **47**(1), 513–539 (1996)

43. Varnum, M.E., Grossmann, I., Kitayama, S., Nisbett, R.E.: The origin of cultural differences in cognition: the social orientation hypothesis. Curr. Dir. Psychol. Sci. **19**(1), 9–13 (2010)

44. Villaronga, E.F., Kieseberg, P., Li, T.: Humans forget, machines remember: artificial intelligence and the right to be forgotten. Comput Law Secur Rev **34**(2), 304–313 (2018)

45. van der Waa, J., Nieuwburg, E., Cremers, A., Neerincx, M.: Evaluating XAI: a comparison of rule-based and example-based explanations. Artif. Intell. **291**, 103404 (2021)

46. Wachter, S., Mittelstadt, B., Russell, C.: Counterfactual explanations without opening the black box: automated decisions and the GDPR. Harv. JL Tech. **31**, 841 (2017)

47. Wu, J.Y., Yang, C.C., Liao, C.H., Nian, M.W.: Analytics 2.0 for precision education. Educ. Technol. Soc. **24**(1), 267–279 (2021)

48. Zhang, B.T.: Cognitive learning and the multimodal memory game: toward human-level machine learning. In: IEEE International Joint Conference on Neural Networks (IEEE World Congress on Computational Intelligence), pp. 3261–3267. IEEE (2008)

Mentalistic Stances Towards AI Systems: Beyond the Intentional Stance

Silvia Larghi[(✉)] [ID] and Edoardo Datteri [ID]

RobotiCSS Lab, Laboratory of Robotics for the Cognitive and Social Sciences, Department of Human Sciences for Education "Riccardo Massa", University of Milano-Bicocca, Piazza Dell'Ateneo Nuovo 1 - 20126, Milan, Italy
s.larghi1@campus.unimib.it, edoardo.datteri@unimib.it

Abstract. Under what circumstances do we attribute a mind to AI systems? And, in this case, how do we think their mind works? Answering these questions is crucial to inform the design of safe and trustable AI, to inform research on the ethical, social and legal issues raised by the increasing presence of AI systems in everyday life and to investigate how they can be used as tools to study human and social cognition. This work proposes a philosophical reflection on the possible structure of people's mental models of AI systems. We distinguish between two possible styles of modeling that people may adopt in everyday contexts. Both involve the attribution of mental states and cognitive abilities to the AI system, even though they differ from one another in some relevant aspects. One modeling style is akin to folk psychology and relies on the attribution of beliefs, desires, and other propositional attitudes to the system. The other, which we will refer to as folk-cognitivist, is more akin to the account of the structure of the mind that characterizes classical cognitive science. These modeling styles correspond to different classes of mentalistic stances that people may adopt when they interact with AI systems in ordinary contexts.

Keywords: Philosophy of artificial intelligence · Philosophy of cognitive science · Human-AI interaction · Mental state attribution

1 Introduction

It has been recently claimed that Generative Artificial Intelligence (AI) systems, such as Open AI's ChatGPT, can be hypnotized [1]. Whether this is the case or not, it is interesting to note that a psychological term denoting a cognitive alteration is used to characterize aspects of the functioning of AI systems of a particular sort. Similarly, situations where the content generated by Large Language Models (LLMs) results "nonsensical or unfaithful to the provided source content" are called hallucinations [2]. Claims of this sort should come as no surprise. Indeed, today's AI systems often display characteristics, like the striking fluidity of natural language interaction and the consistency of the generated texts, which may be expected to induce people to occasionally attribute mental states and other cognitive abilities to them. Consistently, in a growing number

A. Aldini (Ed.): SEFM 2023, LNCS 14568, pp. 28–41, 2024.
https://doi.org/10.1007/978-3-031-66021-4_2

of research studies, paradigms, methods and constructs used to study human psychology (including the theory of mind) are applied to the modeling of LLMs psychology [3–5]. A case in point is Dietz and colleagues [6], who studied adults' and children's understanding of the mind of conversational agents such as smart speakers. Banks [7] presented empirical evidence on humans' theory of robots' mind.

These considerations prompt questions that have an empirical and a philosophical side. Under what circumstances do we attribute a mind to AI systems? And, in this case, how do we think their mind works? Answering these questions is important to inform the design of safe and trustable AI [8, 9], to inform research on the ethical, social and legal issues raised by the increasing presence of AI systems in everyday life—from social robots to conversational agents and other sorts of virtual or embodied agents [10–12], and to investigate how they can be used as tools to study human and social cognition [13, 14].

This work proposes a philosophical reflection on the possible structure of people's mental models of AI systems. More specifically, we distinguish here between two possible styles of modeling that people may adopt in everyday contexts. Both involve the attribution of mental states and cognitive abilities to the AI system, even though they differ from one another in some relevant aspects. One modeling style is akin to folk psychology and relies on our commonsensical concept of mental representation. The other, which we will refer to as *folk-cognitivist*, is more akin to the account of the structure of the mind that characterizes classical cognitive science. The main claim made in this paper is that these modeling styles correspond to different classes of mentalistic stances that people may adopt when they interact with AI systems in ordinary contexts. The analysis presented here is part of a wider research project, whose future step will involve the refinement of the distinctions made here, and the development of theoretical frameworks through which the verbal and non-verbal human-AI interactions can be analyzed, for the purpose of gaining a deeper understanding of people's mental models of AI systems.

The structure of the paper is as follows. The two modeling styles introduced here will be presented in detail in Sect. 2, based on a review of recent studies on the attribution of mental states to artificial systems. Section 3 will provide some examples. Section 4 will delve into the analysis of the two proposed modeling styles, exploring their differences. Section 5 will provide a summary and concluding remarks.

2 Mentalistic Stances Towards AI Systems

2.1 Mental State Attribution to AI Systems

The attribution of mental states to artificial agents has been widely studied in human-robot interaction and human-computer interaction.

Thellman and colleagues [15] review a rich list of studies on mental states attribution to robots carried out from several diverse perspectives, including psychology, neuroscience, computer science, and philosophy. This literature starts from the presupposition that people may adopt either a mentalistic or a non-mentalistic stance toward artificial agents while explaining and predicting their behavior [9]. While mentalistic explanations of behavior refer to the mind and mental capacities of the agent [16], non-mentalistic explanations of behavior do not refer, either implicitly or explicitly, to the systems' mind: they are typically (but not necessarily) based on the theoretical vocabulary of physics and/or electronics (e.g., the robot is stuck because the battery is low).

Notably, the studies published so far on the *mentalistic* explanation of artificial agents' behavior heavily rely on Dennett's conceptual framework [17, 18]. Famously, according to Dennett, there are several possible stances one can adopt to explain and predict the behavior of a system: the physical stance (where one explains the systems' behavior with reference to its physical states), the design stance (which refers to the functional design of the system), and the so-called intentional stance. Adopting the intentional stance towards an artificial agent consists, in Dennett's framework, in explaining and predicting the system's behavior by (a) ascribing beliefs, desires, intentions and other intentional states to it, and (b) assuming that the system will always do what is most rational to do given its current beliefs and desires.

The contemporary literature on the attribution of mental states to artificial agents is highly based on Dennett's intentional stance. Thellman and colleagues [19] investigated whether and how the mechanisms underlying people's judgments of robot behavior overlap with or differ from the case of human or animal behavior, exploring people's intentional stance toward robots versus that toward humans. Marchesi and colleagues [20] investigated the extent to which individuals adopt the intentional stance toward humanoid robots with respect to humans. Thellman and Ziemke [21] provide conceptual considerations regarding the notion of intentional stance versus those of folk psychology and theory of mind often partially overlapping in the contemporary literature on mental state attribution; they also provide considerations on methods for the empirical study of intentional stance toward robots. Various studies are aimed at investigating the factors involved in modulating the adoption of intentional stance toward artificial agents, such as the type of formal education [22] or the agent's movements (in terms of surprising or unsurprising behavior [23]).

Perez Osorio and Wykowska [24] provide a review of the many studies trying to assess whether and in what conditions people adopt the intentional stance towards robotic systems based on empirical tools such as the questionnaire proposed by Marchesi and colleagues [25]. The Intentional Stance Questionnaire proposed in the study by Marchesi and colleagues probes the stance adopted by the participants by contrasting mentalistic explanations of the agent's behavior with so-called mechanistic explanations. Typically, mentalistic explanations of agents' behavior are associated with adopting the intentional stance towards the agent, while 'mechanistic' explanations of agents' behavior are often taken as synonymous of non-mentalistic explanations, and related to the adoption of a non-mentalistic stance (in Dennett's framework, the design stance). In this paper we address the question of whether the intentional stance is the *only* kind of *mentalistic* stance that people may adopt towards artificial systems in ordinary interactions. We believe it is not, as explained in the rest of the paper. In the next paragraphs we argue that mentalistic explanations need not always presuppose the adoption of the intentional stance, as discussed by Dennett, but can be couched in different styles.

2.2 Styles of Mentalistic Modeling and Mentalistic Stances Towards AI Systems

To pave the way for the ensuing discussion, it is important to clarify what is meant here with "stance". We start from the assumption that, when people interact with other agents, they form mental models of them and use these models to explain and predict the agent's behavior. In our perspective, the formulation of a mental model of an AI system (and its explanatory and predictive use) amounts to "taking a stance" towards it. Stances need not be mentalistic: the mental model to which the previous sentence refers is possessed by the human being interacting with the AI system. Some mental models, however, also assign mental states to artificial agents in senses that will be discussed hereafter.

Mental models and beliefs. The notion of mental model has been explored and discussed by several scholars (most notably, Johnson-Laird [26]). Here we will construe this notion along the lines of Achinstein's analysis of theoretical models in physics [27]. According to Achinstein, a theoretical model possesses the following characteristics.

1. It consists in a set of assumptions about an object or a system.
2. It describes a type of object or system "by attributing to it what might be called an inner structure, composition or mechanism, reference to which will explain various properties exhibited by that object or system".
3. It is an "approximation useful for certain purposes", which implies that the same object or system can be modeled in different ways depending on the purpose.

We construe the notion of a "mental model" of an AI system along these lines. More specifically, in the framework proposed here, mental models of AI systems have the following characteristics:

(A) They can be conceived as sets of beliefs, possessed by the modeler, whose contents express a number of assumptions about the AI system.
(B) These assumptions state that the AI system has a particular structure, composition or mechanism. Reference to this structure can be used to explain and predict the behavior of the AI system.
(C) Moreover, mental models capture some aspects and not others of the modeled system.

Concerning point C), whereas Achinstein uses the term "approximation", we prefer to use the terms "abstraction" and "idealization" that are often used in the philosophical literature on models (see Frigg and Nguyen [28]) to refer to the omission of certain aspects and the introduction of falsities, respectively, in the modeling of the target system.

Our conception of mental model refers to the notion of "belief". This notion is used here with the classical meaning of a propositional attitude, defined by having a certain attitude (believing) towards a content expressed by a proposition (for a concise discussion of this classical interpretation, see Crane [29]). If an agent A holds the belief that another agent B is hungry, then A has an attitude (believing) towards a content expressed by the proposition "B is hungry". A may have different attitudes towards the content expressed by the proposition "B is hungry". For example, A might *desire* that B is hungry. Beliefs can play psychological roles [40] influencing the actions of the believer. For example, agent A's belief that agent B is hungry may induce A to feed B.

We are thus claiming that, while interacting with an AI system, the user may form mental models of the system. These mental models can be understood as sets of beliefs held by the user whose content somehow refers to the modeled system. Some of these beliefs will attribute an inner structure, composition, or mechanism to the system. For the sake of generality, let us assume that the content of these beliefs can be represented in a canonical form as "the AI system S has characteristic X".

Styles of mentalistic modeling. Now, one may surely *believe* that the AI system has certain *beliefs*. More specifically, the user's mental model may include beliefs stating that the AI system is itself characterized by the possession of certain beliefs, desires, intentions, or propositional attitudes of various sorts. We will call this style of mental modeling 'folk-psychological'. Classically, folk psychology, or common-sense psychology consists in the attribution of beliefs, desires and other propositional attitudes to other agents, plus law-like generalizations, such as: If someone wants X and holds the belief that the best way to get X is by doing Y then, *ceteris paribus*, the person will do Y (for general discussions on folk psychology, see Ramsey [30], Jackson & Pettit [31], Horgan & Woodward [32], Stich & Ravenscroft [33], Stich [34]). Note that Dennett's intentional stance can be readily accommodated within this framework. According to Dennett, to adopt the intentional stance towards a system consists in "ascribing to the system the possession of certain information and by supposing it to be directed by certain goals, and then by working out the most reasonable or appropriate action on the basis of these ascriptions and suppositions" [17], where the discussion following in Dennett's seminal article makes it clear that the system's possession of certain information can be equated to the possession of certain beliefs and desires.

A long-standing debate in the philosophy of mind and science concerns the distinction between folk psychology and cognitive science (a classical discussion being made by Stich [34]). Whether a deep distinction exists between the two is out of the scope of this paper. We rather claim that the two approaches to the modeling of the mind differ at least *superficially*. Where folk psychology models the mind in terms of propositional attitudes and law-like generalizations among them, *prima facie*, cognitive theories adopt a different theoretical vocabulary, in which the notions of representation and information-processing modules play a central role. In cognitive theories, the mind is modeled in terms of cognitive modules which perform information-processing functions, typically characterized by I/O relationships, where the inputs and the outputs are representations of external or internal states. Functional, information-processing modeling of the mind in cognitive science has been discussed by several scholars, including Bechtel [35], Fodor [36], Pylyshyn [37]. A paradigmatic case in point is Marr's well-known theory of visual perception [38], which models the visual perception capacity in terms of transformations from low-level to higher-level representations of the perceived scene.

Our contention here is that not only cognitive scientists, but also laypeople, may occasionally form mental models of AI systems that ascribe to them not beliefs and rationality, but rather a set of information-processing modules and representational structures. In this case, we say that they form a *folk-cognitivist stance* towards the AI system. This claim will be illustrated with an example in the following section. In Sect. 4, we will elaborate on the distinction between folk psychology and folk cognitivism, and discuss how our proposal relates to the three aforementioned stances discussed by Daniel Dennett.

3   Some Notional Examples

Consider the following scenario. John asks a smart speaker to play Rihanna's "Umbrella", but the system, in response, plays Bob Dylan's "Visions of Johanna". How will the user explain this behavior?

John might adopt what we have called a folk-psychological modeling style. They might hold (in their mental model) the belief that the system believed that the title of Bob Dylan's song was "Umbrella"; or they may traffic at the second-order level and believe that the smart speaker believed (incorrectly) that the user wanted to listen to "Visions of Johanna". A bystander might form a mental model of the smart speaker and believe that the system believes that Rihanna's songs are not that good and that John must change their musical taste. And so on. In all these notional examples, John and the bystander hold a mental model of the smart speaker which is couched in folk-psychological terms. Or, in Dennett's terms, they are taking an intentional stance towards the system.

On the other hand, the user might adopt what we have called a folk-cognitivist modeling style. In this case, the user would believe, for example, that the smart speaker has a certain mechanism for processing the vocal commands of the user, and that something went wrong after John's request for Rihanna's song: the smart speaker did not correctly decode John's audio input. More verbosely, the user's beliefs about the system might have the following contents. The smart speaker has a sensor (a microphone) that stores the raw input audio data (the user's vocal request) in a part of the memory. Then, there is some information-processing module in the system that transforms the input audio data into a textual representation of the vocal command. The textual representation of the command is then sent to other functional modules which perform a web search of the corresponding song; and so on and so forth. For some reason, the raw input data (flowing from John's utterance "Play Rihanna's 'Umbrella'!") were transformed into the text "Visions of Johanna by Bob Dylan". Or, the decoded text was actually "Umbrella by Rihanna", but the web search returned Bob Dylan's song.

Let us examine some characteristics of this model.

First, there is no reason to exclude that John's mental model of the system could take this form. Mental models, in our framework, are sets of beliefs about the system. In this case, John would believe that the smart speaker has a number of characteristics—an inner structure of mechanism.

Second, at least prima facie, this model does not attribute beliefs, desires, intentions to the smart speaker, as folk-psychological explanations would do. Believing that the system is able to process information is different from believing that the system has beliefs, desires, and intentions, to the same extent and in the same way cognitive science theories are different from folk-psychological theories. This is not a folk-psychological mental model of the functioning of the smart speaker. It does not result from taking the intentional stance towards the system. This distinction will be further discussed in the next section.

Third, this model is mentalistic for the same reason cognitive theories are said to model (people's) *minds*. It refers to information-processing modules that process representations (see Fig. 1). With the possible exception of the microphone and the speaker, no reference is made to the physical structures that eventually implement (whatever this means) these information-processing modules. There is substantial disagreement as to what constitutes cognition (see for example Adam and Aizawa [39]), yet if cognition is computation over representational states, then arguably the smart speaker, according to this model, possesses at least one of the important requirements for being a cognitive system. The folk-cognitivist model discussed here models the mind of the smart speaker in the same sense information-processing theories in cognitive science are typically told to model aspects of people's mind.

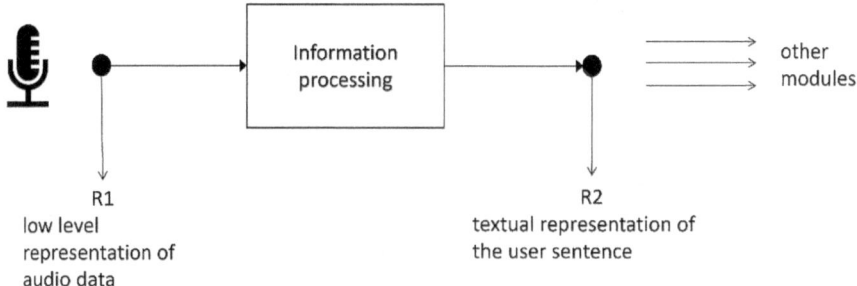

Fig. 1. A possible internal structure attributed to the smart speaker according to the folk-cognitivist modeling style

4   Folk Cognitivism, Intentionality, and the Design Stance

So far, we have argued that a fundamental difference exists between folk-psychological and folk-cognitivist mental models. We have assumed that adopting Dennett's intentional stance is the same thing as adopting a folk-psychological stance toward the system—therefore, that adopting a folk-cognitivist stance is fundamentally different from adopting Dennett's intentional stance. Furthermore, we have suggested that folk-cognitivist models may be called mentalistic, even though they are non-folk-psychological. All these claims, taken together, contribute to the main take-home message of this paper: people can attribute a mind to robots without adopting Dennett's intentional stance. The contemporary literature on mental state attribution to robots exclusively focuses on the intentional stance, and is therefore neglecting another way in which people may model the mind of robots.

We take one of these claims for granted, namely, that Dennett's intentional stance basically consists in adopting a folk-psychological perspective on the system. As already argued, Dennett himself consistently presents the intentional stance as the attribution of beliefs, desires, intentions, and other propositional attitudes to the system. Other claims made here, however, are more controversial and may give rise to questions and objections. In what follows we want to address them, at least provisionally.

(1) Are folk-psychological mental models so different from folk-cognitivist models? In other words, is there a fundamental difference between attributing beliefs and attributing representations and information-processing to the system? One may object that, quite on the contrary, to say that the system possesses a representation whose content is "Visions of Johanna by Bob Dylan" is substantially equivalent to attributing the belief that the user wants to listen to "Visions of Johanna" by Bob Dylan.

(2) Are folk-cognitivist mental models really mentalistic? One may object that having a representation does not imply having a mental state—insofar as this would imply that all the systems to which we commonly attribute representational abilities, such as personal computers, paintings, and underground maps, possess a mind.

(3) Is the folk-cognitivist stance so different from Dennett's design stance? Indeed, Dennett himself occasionally presents the design stance in terms of the attribution of programs, symbols, and computational mechanisms. If this were the case, then it might be objected that we have simply re-invented Dennett's distinction between the intentional and the design stance.

We believe that these possible objections point to aspects of our proposal that need further discussion and refinement. However, the following considerations are meant to provisionally undermine some of these doubts.

4.1 Is Folk Cognitivism so Different from Folk Psychology?

The relation between folk psychology and cognitive science has been extensively debated in the literature by renowned scholars such as Fodor (e.g. [40]) and Stich [34]. These debates stem from the basic consideration that the two styles of behavior explanation differ from each other structurally: whereas folk psychology offers explanations of behavior in terms of generalizations ranging over propositional attitudes, cognitive science explains behavior as produced by information-processing mechanisms. Here we elaborate on this point by emphasizing some of the most prominent differences between the two explanatory approaches.

Firstly, the "atoms" of folk-psychological explanations are propositional attitudes which, as pointed out before, are attitudes towards propositions having certain contents. Propositional attitudes, in a certain sense, *represent* states of affairs (or other propositional attitudes), thus can be conceived as mental representations: if A believes that a cat is on the table, A possesses a mental state that represents a cat on the table. However, William Ramsey [30], among others, has convincingly argued that cognitive science theorizing is based on a notion of representation that is significantly different from folk-psychological representation and is more akin to data structures or maps (see also Cummins [41]). This distinction can be appreciated with reference to desires. In a folk-psychological perspective, one may say that a robotic vacuum cleaner *believes* that the floor is dirty and *desires* to clean it—i.e., it desires to reach a state of affairs in which the floor is clean. One may equate the possession of that belief with the possession of a symbolic representation having the content "the floor is dirty". In other words, one may maintain that (a) saying that the vacuum cleaner has information (a representation) with that content is just the same as (b) saying that it has a belief, thus reducing the difference between the two kinds of "atoms". However, one cannot as easily say the same regarding desires. To say that the vacuum cleaner has a desire whose content is "the floor is clean" is not the same thing as to say that it has a representation or information with the same content. Propositional attitudes (the "atoms" of folk psychology) are in a certain sense "something more", thus structurally different, from the kinds of representations that constitute the building blocks of cognitive theories.

The second difference is closely related to the previous one. Folk-psychological generalizations connect propositional attitudes, whereas cognitivist generalizations are defined in terms of information processing. In a folk-psychological perspective, A will assume that the vacuum cleaner always does what is more rational to do to achieve its current desire (to clean the floor). In a cognitivist perspective—where there are neither beliefs nor desires, but only representations of possible states of affairs—the observer will conjecture that, for example, the system systematically compares a representation of the current state of the floor with the representation of the desired state, and in case of mismatch activates the cleaning device. The two kinds of models postulate different types of generalizations—principles of rationality vs. computational transformations.

This is not to say that the two modeling styles are totally unrelated or irreducible to one another (attempts to bridge folk psychology to cognitive science have been proposed, most notably by Fodor [40]), but that the difference between the two is sufficiently evident to justify empirical investigations of people's folk cognitivist models of robot mind.

4.2   Is Folk Cognitive Science Mentalistic Modeling?

Taking for granted that folk psychology produces mentalistic models and explanations of behavior, one may object that the folk-cognitivist stance is not mentalistic after all: representations—as they are conceived in cognitive science—need not be mentalistic. An immediate response would be that, to support this objection, one must clarify what they mean with "mentalistic model". Sure enough, folk psychology captures some aspects of our everyday, intuitive understanding of the mind. But in some cases it may well go beyond our intuitions. For example, folk-psychological explanations may postulate beliefs, desires, intentions of which the subject *is not aware*—inner representational states that modulate one's behavior with no conscious access. One might wonder why unknowingly having a belief counts as possessing a mental state, while unknowingly possessing a representation of the London underground does not: it is tempting to say that the burden of the proof is on those who make the objection.

More to the point, it is true that representations need not be mentalistic. To be sure, they possess a form of intentionality, insofar as they "are about" things, properties, states of affairs. However, it has been often claimed that a distinction can be made between original (or intrinsic) and derived intentionality, whereas the former only "does not depend on, or derive from, the intentionality of anything else" (Crane [29], p. 29; see also Ramsey [30]). While paintings, statues, words possess derived intentionality, it has been claimed that original intentionality characterizes mental phenomena (this is one aspect of Brentano's thesis about mental phenomena [29, 42]; see also Dennett [18], for a criticism of original intentionality, and Fitch [43] for a discussion of intrinsic aboutness in lower organisms). According to this line of thought, a folk-cognitive stance which attributes representations possessing original intentionality to an artificial system is in a certain sense mentalistic, even though it does not conform to the folk-psychological modeling style. It follows that, in order to determine whether one takes a genuinely folk-cognitive stance towards an artificial system, it is necessary to discover whether they attribute representations in the same (derivate) sense in which a painting represents Napoleon, or in an intrinsic, original sense. Our contention is that, in some cases, people may attribute representational abilities to artificial systems that are not folk-psychological and are embedded with intrinsic intentionality, in the same sense in which cognitive science models aspects of animal mind in terms of information processing.

4.3 Is Folk Cognitive Science Different from the Design Stance?

In our framework, folk-psychological and folk-cognitive models have something in common, as they both characterize the system in terms of intentional entities. However, they considerably differ from one another in the type of entities posited (propositional attitudes and input-output representations): as pointed out before, this makes the two stances significantly different from one another. Here we consider the possible objection that the folk-cognitivist stance is nothing more than Dennett's design stance. Dennett introduces the design stance in his 1971 article in the following terms.

> If one knows exactly how the computer is designed (*including the impermanent part of its design: its program*), one can predict its designed response to any move one makes by following the computation instructions of the program. One's prediction will come true provided only that the computer performs as designed - that is, without breakdown. Different varieties of design-stance predictions can be discerned, but all of them are alike in relying on the notion of function, which is purpose-relative or teleological. That is, a design of a system breaks it up into larger or smaller functional parts, and design-stance predictions are generated by assuming that each functional part will function properly (Dennett [17], p. 87, emphasis added).

This passage—especially the part in italics—may fuel the objection introduced before, as programs prominently figure in cognitive models of the mind. Noteworthy, however, in the same passage Dennett claims that adopting the design stance essentially means characterizing the system in terms of functions and functional parts. And it is far from obvious that the two types of characterizations—in terms of functions and in terms of programs—are one and the same. On the one hand, functional models (see, e.g., Cummins [44], Griffiths [45]) typically decompose the system in terms of components, which are not necessarily localized in physical structures, and characterize them in terms of the function they play in the context of the system. Functional models play a key role in engineering design [46]. Programs, on the other hand, are algorithms written in a programming language. While programs may implement functional models of a system (see, e.g., Durán [47]), it should not be taken for granted that programs are functional models, i.e., that the two modeling styles are equivalent.

Is the folk-cognitivist stance equivalent to the design stance, interpreted as the identification of functional models in the target system? Our provisional claim is that it is a type of design stance, in which functional models are identified in terms of input-output transformations over representations. The notional explanation considered above, for example, identifies information processing modules each performing a particular transformation from input to output representations. One functional model may be devoted, for example, to transforming a digital representation of an auditory input to a digital representation of a text. This is a functional model, as it is identified by reference to the function it performs within the system (speech recognition). However, it is a particular type of functional model, in which the modules are characterized by particular relationships between input and output representations.

In other words, we claim that the folk-cognitivist stance is an intentional (i.e., representational in the sense discussed before) kind of design stance. It coincides neither with the intentional stance nor with the design stance, being a sub-species of the latter one. It is also a sort of mentalistic design stance. While these claims admittedly need further elaboration, we submit that the folk-cognitivist stance represents a peculiar style of modeling that is interestingly different from the stances discussed in the philosophical and psychological so far, and that may be identified in people's explanations of the behavior of technological artifacts.

5 Taking Stocks

In this paper we have distinguished between two mentalistic stances towards the behavior of AI systems. One is called folk-psychological and involves the attribution of rationality and beliefs, desires and other propositional attitudes to the AI system. The other one has been dubbed folk-cognitivist: it is based on the ascription to the AI system of information-processing and representational abilities. We have argued that the two styles of modeling are significantly different from one another, that the folk-cognitivist stance can be construed as mentalistic, and that it is different from the design stance as defined by Dennett. The intentional stance is not the only way to attribute a mind to AI systems: people may also adopt the folk-cognitivist stance.

This consideration gives rise to further questions that call for an empirical analysis of the folk-cognitivist stance. How plausible is it that lay people—in particular, people who are not experts in cognitive science—adopt the folk-cognitivist style? And what methods could be deployed to assess what style people adopt in different situations? As far as the first question is concerned, we note that concepts belonging to computer and cognitive science, like 'information', 'data', 'memory', 'representations' and similar, have become part of our everyday pre-theoretical vocabulary. It would come as no surprise if they were already inflating laypeople's models of AI systems. But this is only a suggestion, that must be evaluated empirically—which leads us to the second question. What methods should be used to study people's mental models of AI systems? Arguably, people's verbal utterances cannot be taken to always express their beliefs literally. John's yelling "You did not understand my request!" does not imply that John literally believes that the system did not understand their request—it could be a metaphorical saying. However, one may devise experimental situations in which folk-psychological and folk-cognitivist models lead to different observable predictions on the behavior of the user. Notably, these predictions might be about the verbal as well as the non-verbal behavior of the user during prolonged interaction with the system. These questions are shaping an ongoing research project that will hopefully lead to a further step towards understanding how people understand AI systems.

Acknowledgements. The authors gratefully acknowledge financial support by the Italian Ministry of University and Research (MUR) in the framework of projects "HERB—Human Explanations of Robotic Behaviour" (PRIN 2022, grant n. 20224X95JC) and "Fostering STEM inclusion among girls with migrant background: the role of educational robots" (PRIN 2022 PNRR, grant n. P2022B2MN2). Parts of this paper have been presented to CRI23 International Conference on Child-Robot Interaction, AIC 2023—9th International Workshop on Artificial Intelligence and Cognition, CIFMA 2023—5th International Workshop on Cognition: Interdisciplinary Foundations, Models and Application, AISC 2023—19th Annual Conference of the Italian Association of Cognitive Sciences. We thank Antonio Lieto and Pierluigi Graziani for their insightful comments.

References

1. Sharma, S.: LLMs like GPT and Bard can be manipulated and hypnotized (2023). https://interestingengineering.com/science/llms-like-gpt-and-bard-can-be-manipulated-and-hypnotized. Accessed 15 Sept 2023
2. Ji, Z., et al.: Survey of hallucination in natural language generation. ACM Comput. Surv. **55**(12), 1–38 (2022)
3. Brunet-Gouet, E., Vidal, N., Roux, P.: Do conversational agents have a theory of mind? A single case study of ChatGPT with the Hinting, False Beliefs and False Photographs, and StrangeStories paradigms. Zenodo (2023)
4. Kosinski, M.: Theory of mind may have spontaneously emerged in large language models (2023). arXiv preprint arXiv:2302.02083
5. Loconte, R., Orrù, G., Tribastone, M., Pietrini, P., Sartori, G.: Challenging ChatGPT 'Intelligence' with human tools: a neuropsychological investigation on prefrontal functioning of a large language model. Intelligence (2023)
6. Dietz, G., Outa, J., Lowe, L., Landay, J. A, Gweon, H.: Theory of AI mind: how adults and children reason about the "mental states" of conversational AI. In: Proceedings of the Annual Meeting of the Cognitive Science Society, vol. 45 (2023)
7. Banks, J.: Theory of mind in social robots: replication of five established human tests. Int. J. Soc. Robot. **12**(2), 403–414 (2020)
8. Ziemke, T.: Understanding robots. Sci. Robot. **5**(46), eabe2987 (2020)
9. De Graaf, M.M., Malle, B.F.: How people explain action (and autonomous intelligent systems should too). In: 2017 AAAI Fall Symposium Series (2017)
10. Carrillo, M.R.: Artificial intelligence: from ethics to law. Telecommun. Policy **44**(6), 101937 (2020)
11. Sullivan, Y.W., Fosso Wamba, S.: Moral judgments in the age of artificial intelligence. J. Bus. Ethics **178**(4), 917–943 (2022)
12. Jaeger, C.B., Levin, D.: If Asimo thinks, does Roomba feel? The legal implications of attributing agency to technology. J. Hum.-Robot Interact. **5**(3), 3–25 (2016)
13. Wykowska, A.: Social robots to test flexibility of human social cognition. Int. J. Soc. Robot. **12**(6), 1203–1211 (2020)

14. Wykowska, A.: Robots as mirrors of the human mind. Curr. Dir. Psychol. Sci. **30**(1), 34–40 (2021)
15. Thellman, S., de Graaf, M., Ziemke, T.: Mental state attribution to robots: a systematic review of conceptions, methods, and findings. ACM Trans. Hum.-Robot Interact. (THRI) **11**(4), 1–51 (2022)
16. De Graaf, M.M., Malle, B.F.: People's explanations of robot behavior subtly reveal mental state inferences. In: 2019 14th ACM/IEEE International Conference on Human-Robot Interaction (HRI), pp. 239–248. IEEE (2019)
17. Dennett, D.C.: Intentional systems. J. Philos. **68**(4), 87–106 (1971)
18. Dennett, D.C.: The Intentional Stance. MIT Press, Cambridge, MA (1987)
19. Thellman, S., Silvervarg, A., Ziemke, T.: Folk-psychological interpretation of human versus humanoid robot behavior: exploring the intentional stance toward robots. Front. Psychol. **8**, 1962 (2017)
20. Marchesi, S., Spatola, N., Perez-Osorio, J., Wykowska, A.: Human versus humanoid. A behavioral investigation of the individual tendency to adopt the intentional stance. In: Proceedings of the 2021 ACM/IEEE International Conference on Human-Robot Interaction, pp. 332–340 (2021)
21. Thellman, S., Ziemke, T.: The intentional stance toward robots: conceptual and methodological considerations. In: The 41st Annual Conference of the Cognitive Science Society, pp. 1097–1103. Cognitive Science Society Inc., Montreal, Canada (2019)
22. Roselli, C., Navare, U.P., Ciardo, F., Wykowska, A.: Type of education affects individuals' adoption of intentional stance towards robots: an EEG study. Int. J. Soc. Robot. 1–12 (2023)
23. Bossema, M., Saunders, R., Allouch, S.B.: Robot body movements and the intentional stance. In: First International Workshop on Designerly HRI Knowledge, IEEE International Conference on Robot and Human Interactive Communication (RO-MAN), vol. 2020 (2020)
24. Perez-Osorio, J., Wykowska, A.: Adopting the intentional stance toward natural and artificial agents. Philos. Psychol. **33**(3), 369–395 (2020)
25. Marchesi, S., Ghiglino, D., Ciardo, F., Perez-Osorio, J., Baykara, E., Wykowska, A.: Do we adopt the intentional stance toward humanoid robots? Front. Psychol. **10**, 450 (2019)
26. Johnson-Laird, P.N.: Mental Models: towards a Cognitive Science of Language, Inference and Consciousness. Cambridge University Press (1983)
27. Achinstein, P.: Theoretical models. Br. J. Philos. Sci. **16**(62), 102–120 (1965)
28. Frigg, R., Nguyen, J.: Models and representation. Springer Handbook of Model-Based Science, pp. 49–102 (2017)
29. Crane, T.: The Mechanical Mind. A Philosophical Introduction to Minds, Machines and Mental Representation, 3rd edn. Routledge (2016)
30. Ramsey, W.M.: Representation Reconsidered. Cambridge University Press (2007)
31. Jackson, F., Pettit, P.: In defence of folk psychology. Philos. Stud. **59**(1), 31–54 (1990)
32. Horgan, T., Woodward, J.: Folk psychology is here to stay. Philos. Rev. **94**(2), 197 (1985)
33. Stich, S., Ravenscroft, I.: What is folk psychology?. In: Collected Papers, vol. 1, pp. 214–234. Oxford University Press (2011)
34. Stich, S.P.: From Folk Psychology to Cognitive Science: the Case Against Belief. The MIT Press (1983)
35. Bechtel, W.: Explanation: mechanism, modularity, and situated cognition. In: The Cambridge Handbook of Situated Cognition, pp. 155–170 (2008)
36. Fodor, J.A.: The Modularity of Mind. The MIT press (1983)
37. Pylyshyn, Z.W.: Computation and Cognition: toward a Foundation for Cognitive Science. MIT Press (1984)
38. Marr, D.: Vision: a Computational Investigation into the Human Representation and Processing of Visual Information. W. H. Freeman, San Francisco (1982)

39. Adams, F., Aizawa, K.: The bounds of cognition. Philosophical Psychology **14**(1), 43–64 (2001)
40. Fodor, J.A.: The Language of Thought. Harvard University Press, Cambridge (Mass.) (1975)
41. Cummins, R.: Meaning and Mental Representation. MIT Press, Cambridge, Mass (1995)
42. Brentano, F.: Psychology from an empirical standpoint. Routledge and Kegan Paul, London (1973), originally published 1874
43. Fitch, W.T.: Nano-intentionality: a defense of intrinsic intentionality. Biol. Philos. **23**(2), 157–177 (2008)
44. Cummins, R.: Functional analysis. J. Philos. **72**(20), 741 (1975)
45. Griffiths, P.E.: Functional analysis and proper functions. Br. J. Philos. Sci. **44**(3), 409–422 (1993)
46. Vermaas, P.E., Kroes, P., Light, A., Moore, S.A. (eds.): Philosophy and Design: from Engineering To Architecture. Springer Science & Business Media (2007)
47. Durán, J.M.: What is a simulation model? Mind. Mach. **30**(3), 301–323 (2020)

Private Memory Confers No Advantage

Samuel A. Alexander[(✉)]

The U.S. Securities and Exchange Commission, New York City, USA
samuelallenalexander@gmail.com

Abstract. Mathematicians and software developers use the word "function" very differently, and yet, sometimes, things that are in practice implemented using the software developer's "function", are mathematically formalized using the mathematician's "function". This mismatch can lead to inaccurate formalisms. We consider a special case of this meta-problem. Various kinds of agents might, in actual practice, make use of private memory, reading and writing to a memory-bank invisible to the ambient environment. In some sense, we humans do this when we silently subvocalize thoughts about the actions we are taking (at least when the environment we're in is too primitive to probe the contents of our brains). Mathematical function formalizations of agents often ignore this ability. We show that in a general agent-environment framework (of which reinforcement learning is a special case), in a technical sense, such private memories do not enable qualitatively different agent behavior.

1 Introduction

There are many different ways to formalize the interaction between an agent and an environment. At a high level, the idea is simple: agent and environment take turns. On the agent's turn, the agent takes an action. On the environment's turn, the environment generates a percept for the agent to see. But precisely how to formalize this depends on various questions about things like what the agent and environment can remember, whether or not agent and/or environment can have an element of randomness, and so on. For example, if we are only interested in memoryless agents who see nothing but the most recent environmental percept, then we can formalize an agent as a function which takes a percept as input and outputs an action (or an action probability distribution). Examples of such agents include the so-called *policies* in the Stable Baselines3 reinforcement learning package [7]. On the other hand, if we are interested in agents who remember the entire interaction so far, then we should formalize an agent as a function which takes a *history* as input, not just a single percept. Examples of such agents include the agents in the Legg-Hutter formalization of reinforcement learning [4,6].

And when formalizing agents in the latter way, there are still questions about what needs to be included in the *history* which is input into the agent-function. For example, if the agent's actions are completely deterministic, then a history need only include the sequence of percepts seen so far. From these, the agent's

A. Aldini (Ed.): SEFM 2023, LNCS 14568, pp. 42–53, 2024.
https://doi.org/10.1007/978-3-031-66021-4_3

actions can be inferred, provided the agent is deterministic. It is superfluous (at least ignoring computational efficiency concerns) for a deterministic agent to remember its own past actions, provided that it remembers the percepts that prompted those actions: the actions themselves, being deterministic, are determined by the percepts. But, if the agent is non-deterministic (i.e., if agents are functions that output action probability distributions rather than just actions), then the history which we input into the agent should include both the past environmental percepts and also the agent's own past actions. Past actions can not, generally, be inferred from past environmental percepts, if said actions had an element of randomness to them.

In addition to memories of past percepts and past actions, we can also imagine agents who have additional *private memory*. For example, we might imagine that the agent has access to a piece of scratch paper stored outside of the environment itself (and thus invisible to the environment) on which the agent can write notes to its own future self. The human hippocampus is not, in reality, an example of such a private memory bank, because, in reality, the human hippocampus is part of the environment (and thus vulnerable to being observed or modified by the outside world). But if, as simplifying assumption, the human agent is modeled as having his brain perfectly isolated, then the human hippocampus does become a place for such private memories to be stored. In such an idealized model, the human agent can silently think to herself things like: "I'm going to see what happens when I push this button. If it hurts me, then I won't push any more buttons like it in the future." And then, later on, the human agent has memory of such internal chatter that she made in the past.

In this paper, we consider agents who have memory of past environmental percepts and of their own past actions. We are concerned with the question: does private memory (in addition to memory of past actions and past environmental percepts) confer any advantage? We will show that the answer is "No," in the following sense. We will show that if π is any agent with private memory in addition to memory of the past agent-environment interaction, then there is an agent π' with only memory of past agent-environment interaction (and no private memory), such that for every finite initial history h of percepts and actions, the probably of h occurring when π interacts with a given environment is exactly equal to the probably of h occurring when π' interacts with that environment. This is trivial if private memories are written deterministically (π' can then reconstruct π's private memories by simulating π on the past percepts and actions which π' remembers). But if private memories have an element of randomness to them, then this result is much less obvious.

Although private memory confers no advantage in the above formal sense, there are other senses in which it might be considered to confer advantage. First of all, the π' described above is not guaranteed to have as good of a runtime complexity as the original π, and indeed, the π' which we will construct in our proof will generally be much more expensive than π. Second, the ability to use private memory might simplify the task of actually implementing a given agent. And thirdly, our proof only applies to traditional environments, not to environments capable of simulating the agent itself, as in [1,3].

2 Formal Definitions

Throughout the paper, we fix a nonempty set $\mathcal{O}$ of *percepts*, a nonempty set $\mathcal{A}$ of *actions*, and a nonempty set $\mathcal{M}$ of *private memories*. We assume $\mathcal{O}$, $\mathcal{A}$, and $\mathcal{M}$ are pairwise disjoint.

Definition 1. If X is any set, by a *probability distribution on X* we mean a function $f : X \to [0,1]$ such that $\sum_{x \in X} f(x) = 1$. For each $x \in X$, we say that f assigns probability $f(x)$ to x. We write $\Delta(X)$ for the set of all probability distributions on X.

2.1 Agents Without Private Memory

Definition 2. *(Histories)*

- By an *environment-seen history*, we mean a sequence $\langle o_1, a_1, \ldots, o_n, a_n \rangle$, where each $o_i \in \mathcal{O}$ and each $a_i \in \mathcal{A}$. We also consider the empty sequence $\langle \rangle$ to be an environment-seen history.
- By an *agent-seen history*, we mean a sequence $\langle o_1, a_1, \ldots, o_{n-1}, a_{n-1}, o_n \rangle$, where each $o_i \in \mathcal{O}$ and each $a_i \in \mathcal{A}$. Note that for each percept $o \in \mathcal{O}$, the length-1 sequence $\langle o \rangle$ is an agent-seen history.
- By a *history* we mean an environment-seen history or an agent-seen history.

Definition 3. *(Environments)* By an *environment* we mean a function μ which takes as input an environment-seen history h, and outputs a probability distribution $\mu(h) \in \Delta(\mathcal{O})$. If h is an environment-seen history and $o \in \mathcal{O}$, we write $\mu(o|h)$ for $(\mu(h))(o)$ (the probability of o according to $\mu(h)$).

Definition 4. *(Deterministic agents without private memory)* By a *deterministic agent without private memory* we mean a function π which takes as input an agent-seen history h, and outputs an action $\pi(h) \in \mathcal{A}$.

Definition 5. *(Non-deterministic agents without private memory)* By a *non-deterministic agent without private memory* we mean a function π which takes as input an agent-seen history h, and outputs a probability distribution $\pi(h) \in \Delta(\mathcal{A})$. If h is an agent-seen history and $a \in \mathcal{A}$, we write $\pi(a|h)$ for $(\pi(h))(a)$ (the probability of a according to $\pi(h)$).

When a deterministic agent π without private memory interacts with an environment μ, we imagine a sequence $o_1, a_1, o_2, a_2, \ldots$ chosen randomly as follows:

- o_1 is chosen randomly using the probability distribution $\mu(\langle \rangle)$.
- $a_1 = \pi(\langle o_1 \rangle)$.
- o_2 is chosen randomly using the probability distribution $\mu(\langle o_1, a_1 \rangle)$.

- $a_2 = \pi(\langle o_1, a_1, o_2 \rangle)$.
- And so on.

When a non-deterministic agent π without private memory interacts with an environment μ, we imagine a sequence $o_1, a_1, o_2, a_2, \ldots$ chosen randomly as follows:

- o_1 is chosen randomly using the probability distribution $\mu(\langle \rangle)$.
- a_1 is chosen randomly using the probability distribution $\pi(\langle o_1 \rangle)$.
- o_2 is chosen randomly using the probability distribution $\mu(\langle o_1, a_1 \rangle)$.
- a_2 is chosen randomly using the probability distribution $\mu(\langle o_1, a_1, o_2 \rangle)$.
- And so on.

But instead of reasoning about these randomly-generated sequences, it is easier to reason about the probability of individual finite histories. In the following definition and throughout the rest of the paper, $\frown$ denotes concatenation.

Definition 6. *(The probability of a history)* Suppose μ is an environment and π is a (deterministic or non-deterministic) agent without private memory. For every history h, let $P_\mu^\pi(h)$ be the probability that h would be an initial sequence of the sequence $o_1, a_1, \ldots$ randomly generated from letting π interact with μ as described above.

Some authors, such as [5], would write $P(h)$ or a variation thereof for $P_\mu^\pi(h)$, if π and μ are clear from context.

2.2   Agents with Private Memory

Definition 7. By an *agent-seen private-memory-augmented history* we mean a sequence $\langle o_1, a_1, m_1, \ldots, o_{n-1}, a_{n-1}, m_{n-1}, o_n \rangle$ where each $o_i \in \mathcal{O}$, each $a_i \in \mathcal{A}$, and each $m_i \in \mathcal{M}$. Note that for each percept $o \in \mathcal{O}$, the length-1 sequence $\langle o \rangle$ is an agent-seen private-memory-augmented history.

Definition 8. *(Deterministic agents with deterministic private memory)* By a *deterministic agent with deterministic private memory* we mean a function π which takes as input a private-memory-augmented agent-seen history h, and outputs an action-memory pair $\pi(h) \in \mathcal{A} \times \mathcal{M}$.

Definition 9. *(Deterministic agents with non-deterministic private memory)* By a *deterministic agent with non-deterministic private memory* we mean a function π which takes as input a private-memory-augmented agent-seen history h, and outputs $\pi(h) \in \mathcal{A} \times \Delta(\mathcal{M})$.

Definition 10. *(Non-deterministic agents with deterministic private memory)* By a *non-deterministic agent with deterministic private memory* we mean a function π which takes as input an agent-seen private-memory-augmented history h, and outputs $\pi(h) \in \Delta(\mathcal{A}) \times \mathcal{M}$.

Definition 11. *(Non-deterministic agents with non-deterministic private memory)* By a *non-deterministic agent with non-deterministic private memory* we mean a function π which takes as input an agent-seen private-memory-augmented history h, and outputs $\pi(h) \in \Delta(\mathcal{A} \times \mathcal{M})$.

When a deterministic agent π with deterministic private memory interacts with an environment μ, we imagine a sequence $o_1, a_1, m_1, o_2, a_2, m_2, \ldots$ chosen randomly as follows:

- o_1 is chosen randomly using the probability distribution $\mu(\langle\rangle)$.
- $(a_1, m_1) = \pi(\langle o_1 \rangle)$.
- o_2 is chosen randomly using the probability distribution $\mu(\langle o_1, a_1 \rangle)$ (note the absense of m_1).
- $(a_2, m_2) = \pi(\langle o_1, a_1, m_1, o_2 \rangle)$.
- o_3 is chosen randomly using the probability distribution $\mu(\langle o_1, a_1, o_2, a_2 \rangle)$ (note the absense of m_1, m_2).
- $(a_3, m_3) = \pi(\langle o_1, a_1, m_1, o_2, a_2, m_2, o_3 \rangle)$.
- And so on.

When a deterministic agent π with non-deterministic private memory interacts with an environment μ, we imagine a sequence $o_1, a_1, m_1, o_2, a_2, m_2, \ldots$ chosen randomly as follows:

- o_1 is chosen randomly using the probability distribution $\mu(\langle\rangle)$.
- $a_1 = a$ and m_1 is chosen randomly using the probability distribution f, where $(a, f) = \pi(\langle o_1 \rangle)$.
- o_2 is chosen randomly using the probability distribution $\mu(\langle o_1, a_1 \rangle)$ (note the absense of m_1).
- $a_2 = a$ and m_2 is chosen randomly using the probability distribution f, where $(a, f) = \pi(\langle o_1, a_1, m_1, o_2 \rangle)$.
- o_3 is chosen randomly using the probability distribution $\mu(\langle o_1, a_1, o_2, a_2 \rangle)$ (note the absense of m_1, m_2).
- $a_3 = a$ and m_3 is chosen randomly using the probability distribution f, where $(a, f) = \pi(\langle o_1, a_1, m_1, o_2, a_2, m_2, o_3 \rangle)$.
- And so on.

When a non-deterministic agent π with deterministic private memory interacts with an environment μ, we imagine a sequence $o_1, a_1, m_1, o_2, a_2, m_2, \ldots$ chosen randomly as follows:

- o_1 is chosen randomly using the probability distribution $\mu(\langle\rangle)$.
- $m_1 = m$ and a_1 is chosen randomly using the probability distribution f, where $(f, m) = \pi(\langle o_1 \rangle)$.
- o_2 is chosen randomly using the probability distribution $\mu(\langle o_1, a_1 \rangle)$ (note the absense of m_1).
- $m_2 = m$ and a_2 is chosen randomly using the probability distribution f, where $(f, m) = \pi(\langle o_1, a_1, m_1, o_2 \rangle)$.

- o_3 is chosen randomly using the probability distribution $\mu(\langle o_1, a_1, o_2, a_2 \rangle)$ (note the absense of m_1, m_2).
- $m_3 = m$ and a_3 is chosen randomly using the probability distribution f, where $(f, m) = \pi(\langle o_1, a_1, m_1, o_2, a_2, m_2, o_3 \rangle)$.
- And so on.

When a non-deterministic agent π with non-deterministic private memory interacts with an environment μ, we imagine a sequence $o_1, a_1, m_1, o_2, a_2, m_2, \ldots$ chosen randomly as follows:

- o_1 is chosen randomly using the probability distribution $\mu(\langle \rangle)$.
- (a_1, m_1) is chosen randomly using the probability distribution $\pi(\langle o_1 \rangle)$.
- o_2 is chosen randomly using the probability distribution $\mu(\langle o_1, a_1 \rangle)$ (note the absense of m_1).
- (a_2, m_2) is chosen randomly using the probability distribution

$$\pi(\langle o_1, a_1, m_1, o_2 \rangle).$$

- o_3 is chosen randomly using the probability distribution $\mu(\langle o_1, a_1, o_2, a_2 \rangle)$ (note the absense of m_1, m_2).
- (a_3, m_3) is chosen randomly using the probability distribution

$$\pi(\langle o_1, a_1, m_1, o_2, a_2, m_2, o_3 \rangle).$$

- And so on.

But instead of reasoning about these randomly-generated sequences, it is easier to reason about the probability of individual finite histories.

2.3   Conditional Probabilities

Definition 12. *(Conditional probabilities of histories)* Suppose μ is an environment, π is an agent (deterministic or non-deterministic, with or without deterministic or non-deterministic private memory), and h is a history.

- Let $P_\mu^\pi(h)$ be the probability that when π interacts with μ, h is an initial segment of the percept-action sequence randomly generated using π and μ as described above.
- Let $P^\pi(h)$ be the probability that h would be an initial segment of the percept-action sequence randomly generated as described above if π were to interact with some environment, on the condition that, in said random generation process, the percepts in h are always selected.
- Let $P_\mu(h)$ be the probability that h would be an initial segment of the percept-action sequence randomly generated as described above if some agent were to interact with μ, on the condition that, in said random generation process, the actions in h are always selected.

Lemma 1. *For any μ, π, h as in Definition 12, $P_\mu^\pi(h) = P^\pi(h) P_\mu(h)$.*

Proof. By the multiplication rule of probability.

3   Private Memory Confers No Advantage

Definition 13. *(Agent equivalence)* Let π be an agent (deterministic or non-deterministic, with or without deterministic or non-deterministic private memory). Let ρ be an agent (deterministic or non-deterministic, with or without deterministic or non-deterministic private memory). We say $\pi \sim \rho$ if the following requirement holds: for every environment μ, for every history h, $P^\pi_\mu(h) = P^\rho_\mu(h)$.

Theorem 1. *(Private memory confers no advantage)*

1. *If π is a deterministic agent with deterministic private memory, there exists a deterministic agent π' without private memory, such that $\pi' \sim \pi$.*
2. *If π is a deterministic agent with non-deterministic private memory, there exists a non-deterministic agent π' without private memory, such that $\pi' \sim \pi$.*
3. *If π is a non-deterministic agent with (deterministic or non-deterministic) private memory, there exists a non-deterministic agent π' without private memory, such that $\pi' \sim \pi$.*

Proof. Fix some $a_0 \in \mathcal{A}$.

First, we will prove (2) and (3) together. Then, we will prove (1).

(2) and (3). Define π' so that for each agent-seen history h and each $a \in \mathcal{A}$,

$$\pi'(a|h) = \begin{cases} \frac{P^\pi(h \frown a)}{P^\pi(h)} & \text{if } P^\pi(h) \neq 0; \\ 1 & \text{if } P^\pi(h) = 0 \text{ and } a = a_0; \\ 0 & \text{if } P^\pi(h) = 0 \text{ and } a \neq a_0. \end{cases}$$

Claim: π' is a non-deterministic agent without private memory. The only thing required to show is that π' outputs probability distributions. Thus, we must show that for any agent-seen history h, $\sum_{a \in \mathcal{A}} \pi'(a|h) = 1$, and for all $a \in \mathcal{A}$, $0 \leq \pi'(a|h) \leq 1$. The latter inequalities are easy to show by induction. To show $\sum_{a \in \mathcal{A}} \pi'(a|h) = 1$, it suffices to consider the case $P^\pi(h) \neq 0$ (the other case is trivial). Suppose a percept-action sequence is randomly generated by letting π interact with some environment, and that the percepts and actions so generated initially match h. The next action must be *some* action in $\mathcal{A}$. Thus $\sum_{a \in \mathcal{A}} P^\pi(h \frown a) = P^\pi(h)$. Therefore

$$\sum_{a \in \mathcal{A}} \pi'(a|h) = \sum_{a \in \mathcal{A}} \frac{P^\pi(h \frown a)}{P^\pi(h)} = \frac{1}{P^\pi(h)} P^\pi(h) = 1,$$

as desired, proving the claim.

It remains to show $\pi' \sim \pi$. Let μ be any environment. We will show by induction on h that for every history h, $P^{\pi'}_\mu(h) = P^\pi_\mu(h)$.

Case 1: $h = \langle \rangle$. Then $P^{\pi'}_\mu(h) = P^\pi_\mu(h) = 1$.

Case 2: $h = h_0 \frown o$ for some $o \in \mathcal{O}$. Then

$$P_\mu^{\pi'}(h) = P_\mu^{\pi'}(h_0 \frown o)$$
$$= P_\mu^{\pi'}(h_0)\mu(o|h_0) \qquad\qquad \text{(Multiplicative rule)}$$
$$= P_\mu^{\pi}(h_0)\mu(o|h_0) \qquad\qquad\quad \text{(Induction)}$$
$$= P_\mu^{\pi}(h_0 \frown o) = P_\mu^{\pi}(h). \qquad \text{(Multiplicative rule)}$$

Case 3: $h = h_0 \frown a$ for some $a \in \mathcal{A}$.

Subcase 3.1: $P^\pi(h_0) = 0$. Then also $P^\pi(h_0 \frown a) = 0$, since one cannot randomly generate initial percepts and actions $h_0 \frown a$ without generating initial percepts and actions h_0 first. Thus $P_\mu^\pi(h_0 \frown a) = 0$ by Lemma 1. Thus

$$P_\mu^{\pi'}(h) = P_\mu^{\pi'}(h_0 \frown a)$$
$$= P_\mu^{\pi'}(h_0)\pi'(a|h_0) \qquad\qquad \text{(Multiplicative rule)}$$
$$= P_\mu^{\pi}(h_0)\pi'(a|h_0) \qquad\qquad\quad \text{(Induction)}$$
$$= 0\pi'(a|h_0) = 0 = P_\mu^{\pi}(h_0 \frown a) = P_\mu^{\pi}(h).$$

Subcase 3.2: $P^\pi(h_0) \neq 0$. Then

$$P_\mu^{\pi'}(h) = P_\mu^{\pi'}(h_0 \frown a)$$
$$= P_\mu^{\pi'}(h_0)\pi'(a|h_0) \qquad\qquad\qquad\qquad\quad \text{(Multiplicative rule)}$$
$$= P_\mu^{\pi}(h_0)\pi'(a|h_0) \qquad\qquad\qquad\qquad\qquad\quad \text{(Induction)}$$
$$= P^\pi(h_0)P_\mu(h_0)\pi'(a|h_0) \qquad\qquad\qquad\qquad \text{(Lemma 1)}$$
$$= P^\pi(h_0)P_\mu(h_0)P^\pi(h_0 \frown a)/P^\pi(h_0) \qquad \text{(Definition of } \pi')$$
$$= P_\mu(h_0)P^\pi(h_0 \frown a) \qquad\qquad\qquad\qquad\qquad \text{(Algebra)}$$
$$= P_\mu(h_0 \frown a)P^\pi(h_0 \frown a) \qquad \text{(Clearly } P_\mu(h_0 \frown a) = P_\mu(h_0))$$
$$= P_\mu^\pi(h_0 \frown a) = P_\mu^\pi(h). \qquad\qquad\qquad\qquad \text{(Lemma 1)}$$

(1) Let π' be the deterministic agent without private memory, defined so that for every agent-seen history h:

- If $P^\pi(h) \neq 0$ then $\pi'(h) = a$ where $(a, m) = \pi(h) \in \mathcal{A} \times \mathcal{M}$.
- If $P^\pi(h) = 0$ then $\pi'(h) = a_0$.

Adopt the following convention: for every action a and agent-seen history h, write $\pi(a|h)$ for 1 if $\pi(h) = a$ or 0 if $\pi(h) \neq a$. Since π is deterministic, clearly if $P^\pi(h) \neq 0$ then $P^\pi(h) = 1$. In that case, if $(a, m) = \pi(h)$, then clearly $P^\pi(h \frown a) = 1$, and by definition $\pi'(h) = a$, so $\pi'(a|h) = 1 = P^\pi(h \frown a)/P^\pi(h)$. On the other hand, if $P^\pi(h) = 0$, then by definition $\pi'(h) = a_0$, so that $\pi'(a|h) = 1$ if $a = a_0$ and $\pi'(a|h) = 0$ if $a \neq a_0$. So altogether,

$$\pi'(a|h) = \begin{cases} \frac{P^\pi(h \frown a)}{P^\pi(h)} & \text{if } P^\pi(h) \neq 0; \\ 1 & \text{if } P^\pi(h) = 0 \text{ and } a = a_0; \\ 0 & \text{if } P^\pi(h) = 0 \text{ and } a \neq a_0, \end{cases}$$

exactly as in the proof of (2) and (3) above. Thus, an identical argument as above shows that $\pi' \sim \pi$.

Note that the π' constructed above is much more computationally expensive than π. If $\mathcal{M}$ is finite, then computing $P^\pi(h)$ (in general) "merely" requires computing π on a number of private-memory-augmented histories which is exponential in the length of h. If $\mathcal{M}$ is infinite, then computing $P^\pi(h)$ (in general) requires computing π infinitely many times and summing infinite series[1].

4   The Reinforcement Learning Special Case and Generalization of Prior Work

One special case of agents and environments is *reinforcement learning* (or RL), in which the percepts generated from an environment include numerical rewards. In RL, agents are considered to perform better or worse depending on their ability or inability to maximize average rewards across the whole space of environments (or some suitable subspace thereof).

At the beginning of Sect. 2 we assumed a nonempty set $\mathcal{O}$ of percepts. All the results of the paper continue to hold if additional structure is imposed on those percepts, e.g., if we further require that each percept include a numerical reward. Thus, this paper's results automatically apply to RL. Let us impose exactly that requirement for the remainder of this section. For each $o \in \mathcal{O}$, let $r(o)$ denote the numerical reward included in o.

The following notation is standard in RL.

Definition 14. If μ is an environment and π is an agent (deterministic or non-deterministic), with or without (deterministic or non-deterministic) private memory, let V_μ^π denote the expected total reward π would obtain from μ (i.e., the expected value of the sum $r(o_1) + r(o_2) + \cdots$ if we generate actions $a_1, a_2, \ldots$, percepts $o_1, o_2, \ldots$, and possibly private memories $m_1, m_2, \ldots$, as in Sect. 2), assuming this expected value exists. If not, V_μ^π is undefined.

For an example in which V_μ^π might be undefined, take μ to be an environment which, every other turn, assigns probability 100% to a percept with reward 1, and every other turn, assigns probability 100% to a percept with reward -1. Any agent interacting with μ would obtain expected total reward $1 - 1 + 1 - 1 + \cdots$, a divergent series.

Proposition 1. *Suppose π, ρ are as in Definition 13. If $\pi \sim \rho$ then for every environment μ, $V_\mu^\pi = V_\mu^\rho$ (and the left-hand side is defined iff the right-hand side is defined).*

---

[1] And if π is sophisticated enough that the machinery on which π is run somehow experiences consciousness when π is computed on various inputs, then there could potentially even be ethical implications about plugging so many inputs into π in order to compute π'. This was perhapse foreshadowed by Nietzsche's doctrine of the eternal return.

Proof. For each agent σ (of whatever kind), let $V^\sigma_{\mu,n}$ denote the expected total reward that σ would obtain after interacting with μ until n percepts are generated. This can be computed by considering all possible histories terminating in an nth percept: for each such history h, multiply the probability $P^\sigma_\mu(h)$ of that history by its total reward $r(h)$ (defined as the sum of $r(o)$ where o ranges over the percepts in h), to obtain the expected total reward contributed by h to $V^\sigma_{\mu,n}$. Thus,

$$V^\pi_{\mu,n} = \sum_h P^\pi_\mu(h)r(h)$$

where h varies over the set of histories terminating in an nth percept. By identical reasoning, $V^\rho_{\mu,h} = \sum_h P^\rho_\mu(h)r(h)$. But $\pi \sim \rho$, so each $P^\pi_\mu(h) = P^\rho_\mu(h)$. Thus each $V^\pi_{\mu,n} = V^\rho_{\mu,n}$. Taking the limit as $n \to \infty$ proves the proposition.

Proposition 1 and Theorem 1 together justify this paper's title: private memory confers no advantage. Not only do equivalent agents have the same probability of resulting in any given history, but if different percepts come with different numerical rewards, then Proposition 1 shows that equivalent agents obtain the same expected total reward from each environment. Thus, as numerically measured by RL rewards, equivalent agents have the same exact numerical performance. Theorem 1 shows that for any agent which makes use of private memories, there is an equivalent agent which does not make use of private memories. Thus, private memories do not enable out-performance of agents without private memories.

This work generalizes the main result of [2]. The authors of that paper considered sequences $\boldsymbol{\pi} = (\pi_1, \ldots, \pi_n)$ of non-deterministic RL agents (without private memory), along with weights $\boldsymbol{w} = (w_1, \ldots, w_n)$ (each $w_i > 0$, with sum $w_1 + \cdots + w_n = 1$). The authors showed that for every such weighted distribution, there is a so-called mixture agent (a non-deterministic agent without private memory) $\boldsymbol{w} \cdot \boldsymbol{\pi}$ with the property that for every environment μ, $V^{\boldsymbol{w} \cdot \boldsymbol{\pi}}_\mu = \boldsymbol{w} \cdot (V^{\pi_1}_\mu, \ldots, V^{\pi_n}_\mu)$. Expressed aphoristically: "the performance of the weighted mixture is the weighted mixture of the performances". But it is trivial to construct such an agent if the agent is allowed to have non-deterministic private memories. Assume $|\mathcal{M}| = n$, say $\mathcal{M} = \{m_1, \ldots, m_n\}$. Let π be the non-deterministic agent, with non-deterministic private memory, defined as follows (where h is a private-memory-augmented history).

- If $h = \langle o \rangle$ then let $\pi(h)$ assign to each pair $(a, m_i) \in \mathcal{A} \times \mathcal{M}$ the probability $\pi(a, m_i | h) = w_i \pi_i(a | h)$. In plain English: with probability w_i, act as π_i and remember (using private memory m_i) having done so.
- Otherwise, h is of the form $(o, a, m_i) \frown h_0$ for some $o \in \mathcal{O}$, $a \in \mathcal{A}$, $m_i \in \mathcal{M}$. Let h^- be the history obtained by deleting all private memories from h. For each pair $(a, m_j) \in \mathcal{A} \times \mathcal{M}$, let $\pi(h)$ assign the probability $\pi(a, m_j | h) = \pi_i(a | h^-)$ if $j = i$, or probability $\pi(a, m_j | h) = 0$ otherwise. In plain English: act as π_i, where i is the agent which you previously remembered (and repeat said memory).

Informally, for its initial turn, π randomly chooses one of the agents $\pi_1, \ldots, \pi_n$ (using the given weights), plays as that agent on that turn, and commits (using private memory) to play as that agent thereafter. Clearly for every environment μ, $V_\mu^\pi = \boldsymbol{w} \cdot (V_\mu^{\pi_1}, \ldots, V_\mu^{\pi_n})$. Proposition 1 and Theorem 1 together show there is a non-deterministic agent π' *without private memory* with the same performance. Thus, the main result from [2] is a special case of this paper.

5   Further Generalization

The results of this paper could be further generalized to agents who, in response to input h:

- (Pre-rationalizing) First generate a memory m (depending on h), and then generate an action which may depend on both h and m.
- (Post-rationalizing) First generate an action a (depending on h), and then generate a memory which may depend on both h and a.
- (Pre-and-post-rationalizing) First generate a memory m_{pre} (depending on h), then an action a (depending on h and m_{pre}), and finally a memory m_{post} (depending on h, m_{pre}, and a).

The detailed formalization, however, becomes very verbose, so we leave it, and the corresponding version of Theorem 1, to the reader.

6   Conclusion

Mathematicians and software developers understand the word "function" in different ways. The software developer's "function" can, in the process of computing an output, do various things which the mathematician's function cannot do, for example, reading from or writing to a persistent memory-bank. We typically use the mathematician's "function" when we formalize certain things, because the mathematician's function is easier to reason about and prove things about. But the mismatch could, a priori, lead to inaccurate formalisms. In this paper, we considered agents, with or without such "private memory", in a very general agent-environment framework, and showed that in a technical sense, every agent with such private memory is equivalent to one without. Thus, in some sense, the software developer's agent's ability to use private memory confers no advantage. This partially justifies formalizing said agents as *mathematical* functions. Of course, software developers' agents can also do various other things that a mathematician's agent cannot do (such as query the system clock, access the computer's camera, or even surf the internet). At least, they can in principle— they might not do so very often in practice. Perhaps it is an advantage of the mathematical formalism that, by formalizing agents as *mathematical* functions, we automatically rule out things like agents who check the system clock or surf the internet in order to compute their outputs. We also rule out private memory, but as this paper shows, that's not a big loss.

Acknowledgements. We gratefully acknowledge Arthur Paul Pedersen, Len Du, Oscar Martinez, and the reviewers for feedback and discussion.

References

1. Alexander, S.A., Castaneda, M., Compher, K., Martinez, O.: Extending environments to measure self-reflection in reinforcement learning. J. Artif. Gen. Intell. **13**(1), 1–24 (2022)
2. Alexander, S.A., Quarel, D., Du, L., Hutter, M.: Universal agent mixtures and the geometry of intelligence. In: AISTATS, PMLR (2023)
3. Bell, J., Linsefors, L., Oesterheld, C., Skalse, J.: Reinforcement learning in Newcomblike environments. In: NeurIPS (2021)
4. Hutter, M.: Universal Artificial Intelligence: sequential Decisions Based on Algorithmic Probability. Springer (2004)
5. Hutter, M.: Discrete MDL predicts in total variation. In: Advances in Neural Information Processing Systems, vol. 22 (2009)
6. Legg, S., Hutter, M.: Universal intelligence: a definition of machine intelligence. Mind. Mach. **17**(4), 391–444 (2007)
7. Raffin, A., Hill, A., Gleave, A., Kanervisto, A., Ernestus, M., Dormann, N.: Stable Baselines3 (2019). https://github.com/DLR-RM/stable-baselines3

Frequentist Probability Logic

Alessandro Aldini⬤, Pierluigi Graziani⬤, and Mirko Tagliaferri$^{(\boxtimes)}$⬤

University of Urbino Carlo Bo, Urbino, Italy
`mirko.tagliaferri@gmail.com,mirko.tagliaferri@uniurb.it`

Abstract. In this paper, we present a logic **FPL** (Frequentist Probability Logic) to reason about probabilities with a relative frequency interpretation. We show that it is possible to interpret the language of **FPL** with the standard semantics for propositional logic. **FPL** can give a peculiar frequentist interpretation of a probability operator. We then give a proof system for the language, prove that the traditional theorems of probability hold, and prove soundness and completeness.

Keywords: Probability logic · Relative frequency · Reasoning about uncertainty

1 Introduction

Probability is important to every human endeavor which involves quantitative data analysis. Moreover, it is safe to claim that probability theory can be considered the best formal tool researchers have to represent uncertainty. From a mathematical standpoint, probability theory is a rigorous and thoroughly studied theory with precise axioms and proven theorems. Thus, it makes sense to try and combine logic and probability, the former providing a qualitative perspective over human reasoning and the latter providing the quantitative analysis [6].

One potential issue with all the formalizations of probability logic is that interpretations of probability are often left tacit; they provide the formal framework and the user applies her preferred interpretation. Although it is true that, as far as an interpretation is consistent with the axioms of probability theory, applying such interpretation to whatever probability logic does not change the characteristics of the logic, it is also true that some interpretations allow for a more fine-grained analysis of why a specific event has a given probability in the first place. For instance, saying that the probability of a fair coin landing heads is $\frac{1}{2}$ is not affected by the interpretation you give to the fraction $\frac{1}{2}$. However, showing that such probability follows from the fact that a coin landed heads twice out of four tosses adds some depth to our understanding of *why* the probability is $\frac{1}{2}$. This last example is an instance of a frequentist interpretation of probability [11] and is extensively employed in scientific settings to establish the probability of different events.

Moreover, a logic that can explicitly capture the idea of relative frequency and that can allow reasoning about probability under said interpretation can become

A. Aldini (Ed.): SEFM 2023, LNCS 14568, pp. 54–71, 2024.
https://doi.org/10.1007/978-3-031-66021-4_4

extremely useful in context of statistical and approximate model checking [10,12] and of probabilistic verification of machine learning [9,13,14].

Statistical model checking refers to techniques that are employed in computer science to check the properties of stochastic systems. Differently from Boolean model checkers, which establish whether a property is reached or not in a model of a system, statistical model checkers start from the assumption that exhaustively exploring all paths of a model might not always be feasible and, thus, quantitative questions should be asked rather than qualitative one. For example, a statistical model checker would look for answers to questions such as: is the probability that the system satisfies a certain property greater than a certain threshold?

In a similar way, the logic proposed here could be employed to construct sophisticated reinforcement learning strategies for machine learning. Reinforcement learning is based on the idea that whenever a classifier gets an answer wrong, it gets punished, while it gets a prize whenever the classification is correct. However, we might want to cluster sets of answers together and punish only classifications that fall below certain thresholds that we have chosen as appropriate. This might thus allow those systems to create appropriate strategies that, although not perfect, might still be useful in contexts in which fuzzy evaluations are sufficient to make the best decision.

The logic presented in this paper allows computer scientists to represent and reason in those contexts with ease and thus represent valuable tools in the hands of the right modeler. The aim of the paper is to enrich the semantics of standard probability logic by capturing explicitly a frequentist interpretation of probability. In particular, we will start from Fagin et al. [7] Probability Logic and apply to it a frequentist semantics. Moreover, we will enrich the language by allowing combinations of propositional and probabilistic formulas, thus allowing an agent to reason about the relationship between local and global results (Sect. 3). We will then show that all meta-theoretic results from [7] hold for our semantic structures (Sect. 4) and we will discuss some future works that make our semantics an interesting evolution which might be of interest to computer scientists (Sect. 5).

2   Related Works

To the best of our knowledge, the works that mostly resemble our approach are those on counting propositions [4] and those on graded modal logics [8]. The former family of works present formal semantics to reason about how many times a specific proposition is true in a model. The general idea presented in those works is really similar to ours, with the only distinction being the lack of connections to probability and its theorems. In this paper, we extend the analysis carried out by those authors by separating the truth of a proposition from its probability (thus capturing an intrinsic notion of uncertainty) and by constructing all the bridges between such formalism and probability logic. On the other hand, graded modal logic offers a way to quantify the uncertainty about the application of specific modal operators to formulas. When such modal operators are interpreted to stand from probability, the connection with our proposal becomes obvious. However, in graded modal logics a relational structure must be assumed, which is not strictly necessary in our semantics. In particular, we claim that graded modal logics might offer a subjective interpretation (due to the accessibility relations) of an objective interpretation of probability (due to the counting of true instances of a proposition in the structure). We offer a completely objective interpretation, creating a stronger tie with probability theory and the reasoning thereof.

3   Frequentist Probability Logic

3.1   Syntax

Take a countable set of primitive atomic formulas At ranging over p_i, with $i \in \mathbb{N}$. We will also assume that a, b range over the set $\mathbb{Z}$.

Definition 1. (*Language of* **FPL**) The language $\mathcal{L}_{\textbf{FPL}}$ ranging over χ_i, with $i \in \mathbb{N}$ is generated recursively by the following three-level grammar, with $\odot \in \{\geq, \leq\}$ and $\circ \in \{+, -\}$:

$$\chi := \varphi \mid \psi \mid \neg\chi \mid \chi \wedge \chi$$
$$\psi := P(\varphi) \odot b \mid P(\varphi) \circ P(\varphi) \odot b$$
$$\varphi := \top \mid p_i \mid \neg\varphi \mid \varphi \wedge \varphi$$

The elements of $\mathcal{L}_{\textbf{FPL}}$ are called formulas. Specifically, the φ_is will be called *propositional formulas*, while the ψ_is will be called *probabilistic formulas*. Finally, the χ_is will be simply called *formulas*.

The following definitions will be employed throughout the paper:

Definition 2. (*Defined formulas*)

- $\bot := \neg\top$;
- $\varphi_i \vee \varphi_j := \neg(\neg\varphi_i \wedge \neg\varphi_j)$;
- $\varphi_i \rightarrow \varphi_j := \neg\varphi_i \vee \varphi_j$;
- $\varphi_i \leftrightarrow \varphi_j := (\varphi_i \rightarrow \varphi_j) \wedge (\varphi_j \rightarrow \varphi_i)$;[1]
- $P(\varphi_i) \odot P(\varphi_j) := (P(\varphi_i) - P(\varphi_j)) \odot 0$;
- $P(\varphi) = b := P(\varphi) \geq b \wedge P(\varphi) \leq b$;
- $P(\varphi) > b := \neg(P(\varphi) \leq b)$;
- $P(\varphi) < b := \neg(P(\varphi) \geq b)$.

Example 1. Imagine that Alice wants to draw some conclusions about a Monte Carlo simulation she ran over throwing two dice. For the purpose of this example, she will only focus on the outcomes of the dice that correspond to three (it is not hard to expand the example, but to keep it simple, we will limit ourselves to the case mentioned). Thus, the important propositions that Alice is interested

---

[1] The same abbreviations will also be employed for the χ_i formulas.

in are p_{13} which stands for "the outcome of the throw of die 1 is 3", and p_{23} which stands for "the outcome of the throw of die 2 is 3". Inside **FPL**, Alice could then reason about formulas such as "both dice landed on the value 3" $(p_{13} \wedge p_{23})$; "whenever die one landed on three, die two did not" $(p_{13} \rightarrow \neg p_{23})$; "the relative frequency of die one landing three is $\frac{1}{6}$" $(P(p_{13}) = \frac{1}{6})$; "the relative frequency of die one landing three is lower than $\frac{1}{10}$" $(P(p_{13}) < \frac{1}{10})$.

3.2   Semantics

To define the semantics of **FPL**, we first introduce Frequentist Models.

Definition 3. (*Frequentist Models*) A *frequentist model* is a couple $\mathfrak{M} = (O, v)$, where O is a non-empty finite set of possible outcomes ranging over o_i with $i \in \mathbb{N}$, and v is a valuation function $v : O \times At \rightarrow \{0, 1\}$.

Intuitively, an outcome $o_i \in O$ can be seen as a valuation of all the atomic formulas of the language. Those outcomes could also be interpreted as experiments (or Montecarlo simulations) in which the atomic formulas are tested in order to verify whether they hold or not. The valuation function, on the other hand, indicates the results of the evaluations/experiment, indicating which formulas turned out to be true.

Example 2. We continue from Example 1. In order to evaluate the formulas Alice is interested in, she can perform a Monte Carlo simulation, throwing the two dice repeatedly and keeping track of the results. Imagine that Alice threw the two dice 30 times. The 30 throws performed by Alice would constitute the set O of possible outcomes. Moreover, the valuation function v would be equivalent to the notes that Alice made about the specific outcomes of the throws. Imagine that out of the 30 throws, die one landed on three five times (e.g., on throws one, three, eight, fourteen, twenty-seven), while die two landed on three on ten different occasions (e.g., on throws two, four, six, eleven, fourteen, sixteen, twenty-two, twenty-three, twenty-eight, and thirty). Formally, this would be captured by a model $\mathfrak{M}$ where $O = \{o_i \mid 1 \leq i \leq 30\}$, and v would be defined as follows (we will only specify the cases in which the formulas are true):

- $v(o_1, p_{13}) = v(o_3, p_{13}) = v(o_8, p_{13}) = v(o_{14}, p_{13}) = v(o_{27}, p_{13}) = 1$;
- $v(o_2, p_{23}) = v(o_4, p_{23}) = v(o_6, p_{23}) = v(o_{11}, p_{23}) = v(o_{14}, p_{23}) = v(o_{16}, p_{23}) = v(o_{22}, p_{23}) = v(o_{23}, p_{23}) = v(o_{28}, p_{23}) = v(o_{30}, p_{23}) = 1$.

It is fairly easy to extend the valuation function to all propositional formulas φ through induction over the structure of the formulas:

Definition 4. (*Extended Valuation*) Given a frequentist model $\mathfrak{M}$, the valuation function v is extended to all the propositional formulas recursively as follows (we use v^e to indicate the extension of v):

- $v^e(o_i, p_j) = v(o_i, p_j)$;
- $v^e(o_i, \neg \varphi) = 1$ iff $v^e(o_i, \varphi) = 0$;
- $v^e(o_i, \varphi_j \wedge \varphi_m) = 1$ iff $v^e(o_i, \varphi_j) = 1$ and $v^e(o_i, \varphi_m) = 1$.

Valuation functions do not apply to probability formulas. Since we are not allowing the nesting of probability operators, this will not constitute a problem.

Definition 5. (*Validating Sets*) Given a frequentist model $\mathfrak{M}$, the *validating set* of a propositional formula φ, indicated with $[\![\varphi]\!]$, is a subset of O ($[\![\varphi]\!] \subseteq O$) s.t. $o_i \in [\![\varphi]\!]$ iff $v^e(o_i, \varphi) = 1$.

Example 3. Given our example, the validating sets for propositions p_{13} and p_{23} would be, respectively:

- $[\![p_{13}]\!] = (o_1, o_3, o_8, o_{14}, o_{27})$;
- $[\![p_{23}]\!] = (o_2, o_4, o_6, o_{11}, o_{14}, o_{16}, o_{22}, o_{23}, o_{28}, o_{30})$.

The validating set of a formula, sometimes referred to as the truth set of the formula, simply indicates the set of outcomes where the formula is true.

Proposition 1 (Properties of Validating Sets). *Given a validating set $[\![\varphi]\!]$, the following properties follow:*

- $[\![\neg\varphi]\!] = O \setminus [\![\varphi]\!]$;
- $[\![\varphi_j \wedge \varphi_m]\!] = [\![\varphi_j]\!] \cap [\![\varphi_m]\!]$.

Proof. The proof follows directly from Definitions 4 and 5.

Definition 6. (*Validating Space*) The **validating space** $\mathcal{V}_{\mathfrak{M}}$ of a model $\mathfrak{M}$, is the set of all validating sets of propositional formulas of the language $\mathcal{L}_{\mathbf{FPL}}$:

$$\mathcal{V}_{\mathfrak{M}} = \{[\![\varphi]\!] \mid \varphi \in \mathcal{L}_{\mathbf{FPL}}\} \tag{1}$$

Proposition 2. *Given a frequentist model $\mathfrak{M}$, the validating space $\mathcal{V}_{\mathfrak{M}}$ forms a σ-algebra over $O \in \mathfrak{M}$.*

Proof. Take an arbitrary model $\mathfrak{M} = (O, v)$. First note that by Definitions 6 and 5, it follows that $\mathcal{V}_{\mathfrak{M}}$ is a set of subsets of O. Since $\top \in \mathcal{L}_{\mathbf{FPL}}$ and $[\![\top]\!] = O$, it follows that $O \in \mathcal{V}_{\mathfrak{M}}$.

Assume that $[\![\varphi]\!] \in \mathcal{V}_{\mathfrak{M}}$. Since $\mathcal{L}_{\mathbf{FPL}}$ is closed under negation, i.e., if $\varphi \in \mathcal{L}_{\mathbf{FPL}}$, then $\neg\varphi \in \mathcal{L}_{\mathbf{FPL}}$, it follows that $[\![\neg\varphi]\!] \in \mathcal{V}_{\mathfrak{M}}$. By Proposition 1, $[\![\neg\varphi]\!] = O \setminus [\![\varphi]\!]$, which is the complement of $[\![\varphi]\!]$. Therefore, $\mathcal{V}_{\mathfrak{M}}$ is closed under complement.

Assume that $[\![\varphi_i]\!] \in \mathcal{V}_{\mathfrak{M}}$ and that $[\![\varphi_j]\!] \in \mathcal{V}_{\mathfrak{M}}$. Since $\mathcal{L}_{\mathbf{FPL}}$ is closed under disjunction (through Definition 2), i.e., if $\varphi_i \in \mathcal{L}_{\mathbf{FPL}}$ and $\varphi_j \in \mathcal{L}_{\mathbf{FPL}}$, then $\varphi_i \vee \varphi_j \in \mathcal{L}_{\mathbf{FPL}}$, it follows that $[\![\varphi_i \vee \varphi_j]\!] \in \mathcal{V}_{\mathfrak{M}}$. Through set operations and Proposition 1, $[\![\varphi_i \vee \varphi_j]\!] = [\![\varphi_i]\!] \cup [\![\varphi_j]\!]$. Therefore, $\mathcal{V}_{\mathfrak{M}}$ is closed under finite union.

From the previous four properties, it follows that $\mathcal{V}_{\mathfrak{M}}$ forms an algebra over $O \in \mathfrak{M}$.

Moreover, since O is finite, $\mathcal{V}_{\mathfrak{M}}$ is also a σ-algebra over O, since countable union collapses onto finite union.

In the following definitions, we will indicate with $|O|$, and $|[\![\varphi]\!]|$ the cardinality of, respectively, the outcomes set and the validating set.

Definition 7. (*Relative Frequency Function*) Given a model $\mathfrak{M}$, it is possible to define a **relative frequency function** τ that assigns to every propositional formula of the language its relative frequency. The assignment procedure for τ is defined as follows: $\tau(\varphi) = \frac{\|[\varphi]\|}{|O|}$.

Definition 8. (*Probability Measure*) Given a set S of states and a σ-algebra $\mathcal{X}$ of measurable sets X_i, a **probability measure** is a function μ which takes as arguments the measurable sets and returns as values a number from $[0,1] \in \mathbb{Q}$. Moreover, a probability measure satisfies the following properties:

- $\mu(X_i) \geq 0$, for all $X_i \in \mathcal{X}$;
- $\mu(S) = 1$;
- If $X_i \cap X_j = \emptyset$, then $\mu(X_i \cup X_j) = \mu(X_i) + \mu(X_j)$.

The triple $(S, \mathcal{X}, \mu)$ is called a **probability space**.

Remark 1. Given a frequentist model $\mathfrak{M}$, the triple $(O, \mathcal{V}_{\mathfrak{M}}, \tau)$ is a probability space.

In the semantics of **FPL**, formulas are interpreted over frequentist models. Propositional formulas will be interpreted locally, while probability formulas will be interpreted globally over the whole model. In particular, the semantics of the operator $P(\cdot)$ is given in terms of the relative frequency function $\tau(\cdot)$.

Definition 9. (*Truth of propositional formulas*) Let $\varphi \in \mathcal{L}_{\textbf{FPL}}$ be a propositional formula and $\mathfrak{M} = (O, v)$ be a frequentist model. We inductively define the notion of φ being verified (or satisfied) by an outcome $o_i \in O$ in $\mathfrak{M}$, written $o_i \models_{\mathfrak{M}} \varphi$, as follows:

1. $o_i \models_{\mathfrak{M}} \top$, always;
2. $o_i \models_{\mathfrak{M}} p_j$ iff $v^e(o_i, p_j) = 1$;
3. $o_i \models_{\mathfrak{M}} \neg\varphi$ iff $o_i \nvDash_{\mathfrak{M}} \varphi$;
4. $o_i \models_{\mathfrak{M}} \varphi_j \wedge \varphi_m$ iff $o_i \models_{\mathfrak{M}} \varphi_j$ and $o_i \models_{\mathfrak{M}} \varphi_m$.

Definition 10. (*Truth of probability formulas*) Let $\psi \in \mathcal{L}_{\textbf{FPL}}$ be a probability formula, $\chi \in \mathcal{L}_{\textbf{FPL}}$ a general formula, and $\mathfrak{M} = (O, v)$ be a frequentist model. We inductively define the notion of ψ (or χ) being verified (or satisfied) by an outcome $o_i \in O$ in $\mathfrak{M}$, written $o_i \models_{\mathfrak{M}} \psi$ (the same notation applies to χ), as follows:

5. $o_i \models_{\mathfrak{M}} P(\varphi) \odot b$ iff $\tau(\varphi) \odot b$;
6. $o_i \models_{\mathfrak{M}} P(\varphi_j) \circ P(\varphi_m) \odot b$ iff $\tau(\varphi_j) \circ \tau(\varphi_m) \odot b$;
7. $o_i \models_{\mathfrak{M}} \neg\chi$ iff $o_i \nvDash_{\mathfrak{M}} \chi$;
8. $o_i \models_{\mathfrak{M}} \chi_j \wedge \chi_m$ iff $o_i \models_{\mathfrak{M}} \chi_j$ and $o_i \models_{\mathfrak{M}} \chi_m$.[2]

---

[2] Note that if χ has no occurrences of probability formulas inside of it, then the truth definition collapses onto that of propositional formulas.

The semantic interpretation of the operation signs + (addition) and − (subtraction) is the standard one from arithmetic, as is that of the inequality/equality signs. It is easy to notice that in the truth definition of probability formulas, the specific outcome chosen plays no role (i.e., there is no difference in evaluating a probability formula ψ in o_1 or in o_2, whatever the outcomes say). This is perfectly reasonable since we already said that the probability formulas are evaluated globally, rather than locally. Therefore, whenever a probability formula is satisfied at a pointed frequentist model, it is satisfied in all the models. We kept the notation of pointed models for probability formulas for the coherence of notation and to more easily transition to general formulas χs. Note that for a χ formula, the valuation must be taken at a pointed model, since they might contain propositional formulas as their elements, which would then have to be evaluated locally.

Remark 2. Note that by simply dropping the probabilistic formulas from the language, the resulting language is that of propositional logic. Indeed, the semantics is also that of propositional logic, where the couples (O, v) can be seen as truth assignment to atomic formulas. This implies that, strictly speaking, **FPL** expands the expressivity of propositional logic, maintaining the same semantics elements of it.

Example 4. We are now in a position to show how Alice could evaluate her formulas. Recall that Alice was interested in those propositions: "both dice landed on the value 3" ($p_{1_3} \wedge p_{2_3}$); "whenever die one landed on three, dice two did not" ($p_{1_3} \rightarrow \neg p_{2_3}$); "the relative frequency of die one landing three is $\frac{1}{6}$" ($P(p_{1_3}) = \frac{1}{6}$); "the relative frequency of die one landing three is lower than $\frac{1}{10}$" ($P(p_{1_3}) < \frac{1}{10}$). As should be expected, for the propositional propositions that Alice is asking, it must be specified which specific trial she has in mind. However, for the probability propositions, her evaluations should be global, again, as expected. In particular, in **FPL**, it is easy to show that $p_{1_3} \wedge p_{2_3}$ is satisfied only by outcome o_{14}, i.e., $o_{14} \models_{\mathfrak{M}} p_{1_3} \wedge p_{2_3}$. At the same time, it is always false that $p_{1_3} \rightarrow \neg p_{2_3}$. Moreover, it happens to be true in the model we constructed that $P(p_{1_3}) = \frac{1}{6}$, i.e., $\models_{\mathfrak{M}} P(p_{1_3}) = \frac{1}{6}$, while it is false that $P(p_{1_3}) < \frac{1}{10}$.

Definition 11. (*Validity of propositional and probability formulas*) A propositional formula φ is said to be **satisfiable in a frequentist model** $\mathfrak{M} = (O, v)$ iff $\exists o_i \in O$ such that $o_i \models_{\mathfrak{M}} \varphi$. A propositional formula φ is said to be **valid in a frequentist model** $\mathfrak{M} = (O, v)$ (in symbols, $\models_{\mathfrak{M}} \varphi$) iff $\forall o_i \in O \in \mathfrak{M}, o_i \models_{\mathfrak{M}} \varphi$. For a probability formula, as already mentioned, satisfiability and validity in a model collapse, since the valuation of a probability formula ψ is global on the model. Thus, a probability formula ψ is said to be **valid in a frequentist model** $\mathfrak{M} = (O, v)$ iff $\models_{\mathfrak{M}} \psi$. A formula χ retains the same definitions of a propositional formula, with the distinction that whenever χ only contains probability formulas, then satisfiability collapses onto validity in a model.

Proposition 3 (Relation between validity and satisfiability). *A propositional formula φ is valid iff its negation is not satisfiable in a frequentist model.*

Moreover, a probability formula ψ is valid iff its negation is not valid in a frequentist model.

Proof. Assume that a propositional formula φ is valid. This is equivalent to the fact that the formula is satisfied in every outcome o_i of every model $\mathfrak{M}$. Take an arbitrary model $\mathfrak{M}$ and an arbitrary outcome $o_i \in O \in \mathfrak{M}$. Take the negation of φ, i.e., $\neg\varphi$. To satisfy such formula, it must hold that $o_i \models_{\mathfrak{M}} \neg\varphi$, which, by definition, means that $o_i \not\models_{\mathfrak{M}} \varphi$. However, by assumption, we know that there is no o_i s.t. $o_i \not\models_{\mathfrak{M}} \varphi$, thus there is no o_i s.t. $o_i \models_{\mathfrak{M}} \neg\varphi$.

For the probability formulas, the argument is similar, with the difference being that validity in a model is taken into consideration instead of satisfiability. This is due to the fact that probability formulas are evaluated globally on the model and not at a specific outcome.

An example of a valid formula in **FPL** would be $P(\varphi) \geq \frac{1}{2} \rightarrow P(\varphi) \geq \frac{1}{4}$.

Proposition 4. *The following two sentences are equivalent, where φ is a propositional formula:*

(a) $\models \varphi$;
(b) $\models P(\varphi) = 1$.

Proof. Take an arbitrary model $\mathfrak{M} = (O, v)$.

(From (a) to (b)). Assume that $\models \varphi$. It therefore follows, by Definition 11, that $\forall o_i \in O, o_i \models_{\mathfrak{M}} \varphi$. By Definition 9, that $\forall o_i \in O, v^e(o_i, \varphi) = 1$. The validating set $[\![\varphi]\!]$ is therefore equivalent to O, from which it follows that $\frac{|[\![\varphi]\!]|}{|O|} = 1$, which in turn implies $\frac{|[\![\varphi]\!]|}{|O|} = 1$. This is the definition of $\models_{\mathfrak{M}} P(\varphi) = 1$, via Definition 10. Since the model $\mathfrak{M}$ was taken arbitrarily, this holds for all possible models chosen.[3]

(From (b) to (a)). Assume that $\models_{\mathfrak{M}} P(\varphi) = 1$. It follows, by Definition 10, that $\frac{|[\![\varphi]\!]|}{|O|} = 1$, i.e., $|[\![\varphi]\!]| = |O|$. By Definition 5, $[\![\varphi]\!] \subseteq O$. Since O is finite, the only possible case in which the two conditions $[\![\varphi]\!] \subseteq O$ and $|[\![\varphi]\!]| = |O|$ are both true at the same time is when $[\![\varphi]\!] = O$. From such equality and Definition 5, it follows that $\forall o_i \in O, v^e(o_i, \varphi) = 1$, which, by Definition 9, means that $\forall o_i \in O, o_i \models_{\mathfrak{M}} \varphi$. Finally, by Definition 11, this means that $\models_{\mathfrak{M}} \varphi$. Since the model $\mathfrak{M}$ was taken arbitrarily, this holds for all possible models chosen.[4]

3.3   Axiomatic System for FPL

We will now provide an axiomatic system for **FPL**. We will then show that such an axiomatic system is sound and complete with respect to frequentist models.

---

[3] Technically, both $P(\varphi)$ and 1 should be qualified with a coefficient a. However, multiplying both for the same number, would not change the result of the valuation. Thus the formula would still be valid for any a chosen.

[4] Note that it is assumed in **FPL** that O is finite. While using an infinite O would not change the (a) to (b) it might indeed make the (b) to (a) direction false.

Definition 12. (*Frequentist Probability Logic*) A **frequentist probability logic** (FPL) is a set $\Lambda \subseteq \mathcal{L}_{\textbf{FPL}}$ containing (i) all propositional tautologies, (ii) all the substitution instances of valid formulas about linear inequalities, (iii) all the substitution instances of the following axiom schemata, for $\varphi, \psi \in \mathcal{L}_{\textbf{FPL}}$:

- Ax. 1 $P(\varphi) \geq 0$;
- Ax. 2 $P(\top) = 1$;
- Ax. 3 If $\neg(\varphi_i \wedge \varphi_j)$, then $P(\varphi_i \vee \varphi_j) = P(\varphi_i) + P(\varphi_j)$;[5]
- Ax. 4 If $(\varphi_i \leftrightarrow \varphi_j)$, then $P(\varphi_i) = P(\varphi_j)$.

Λ is also closed under Modus Ponens (MP).

Instances of (i) formalize reasoning about propositional formulas and Boolean combinations between formulas (either propositional or probability formulas or a mix of the two); instances of (ii) formalize reasoning about formulas of the form $P(\varphi_i) \circ P(\varphi_j) \odot b$, where the $P(\varphi)$s can be seen as variables in the inequalities. A formula about linear inequality is valid only if every numerical assignment to the variables in the inequality. i.e., the $\tau(\varphi)$ of the $P(\varphi)$s appearing in the inequality, satisfy the inequality. Ax. 1 states that probability formulas can never receive a negative value. Ax. 2 states that the probability of a tautology is always 1 (see also Proposition 4). Ax. 3 captures the idea of finite additivity of probability theory; this axiom is also what creates the connection between $P(\varphi) \circ P(\varphi) \odot b$ formulas and $P(\varphi) \odot b$ formulas (i.e., it is an interaction axiom between simple probability formulas and linear combinations of them). Ax. 4 captures the idea that equivalent formulas should have the same probability.

Definition 13. We say that a formula χ is **provable** in Λ ($\vdash_\Lambda \chi$) iff the formula is an instance of an axiom schema in Λ or is obtained through the application of MP to formulas already proven. A formula χ is **derivable** in Λ from a set of premises Γ ($\Gamma \vdash_\Lambda \chi$) iff the formula is an instance of an axiom schema in Λ, is an instance of a formula contained in Γ, or is obtained through the application of MP to formulas already derived or proven.

We say that a formula χ is **inconsistent** if its negation $\neg\chi$ is provable ($\vdash_\Lambda \neg\chi$). A formula is **consistent** otherwise.

3.4   Probability Theorems

We will now show that the main theorems from probability theory can be proven in **FPL**. We will provide semantic proofs rather than syntactic ones. Our choice is justified by the fact that the major novel contribution to the literature of **FPL** is semantic in nature, rather than syntactic. For convenience, we also provide syntactic proofs in the appendix of this paper.

Theorem 1. (Probability of $\bot$) $\models P(\bot) = 0$

---

[5] Finite additivity could also be expressed by the formula $P(\varphi_i) = P(\varphi_i \wedge \varphi_j) + P(\varphi_i \wedge \neg\varphi_j)$. This fact is important since we will use this formulation in our proof of completeness.

Proof. Take an arbitrary model $\mathfrak{M}$ and an arbitrary outcome $o_i \in O \in \mathfrak{M}$. By Definition 2, $\bot := \neg\top$, and by Definition 9, $\top$ holds for every $o_i \in O \in \mathfrak{M}$. By Definition 5, it follows that $[\![\top]\!] = O$. By Definition 1, it follows that $[\![\bot]\!] = O \setminus [\![\top]\!]$. By substitution, it follows that $[\![\bot]\!] = \emptyset$. Therefore, $\tau(\bot) = \frac{|\emptyset|}{|O|}$. Whatever the cardinality of O is, the cardinality of $\emptyset$ is zero. It follows that $\tau(\bot) = 0$, which means that $P(\bot) = 0$ holds. Since the outcome and the model were chosen arbitrarily, the results holds for any outcome and any model.

Theorem 2 (Finite Additivity).

$$If \models \neg(\varphi_i \wedge \varphi_j), then \models P(\varphi_i \vee \varphi_j) = P(\varphi_i) + P(\varphi_j)$$

Proof. Take an arbitrary model $\mathfrak{M}$. Assume that $\models \neg(\varphi_i \wedge \varphi_j)$, thus, in particular, $\models_{\mathfrak{M}} \neg(\varphi_i \wedge \varphi_j)$. This means that for each $o_i \in O \in \mathfrak{M}$, either $o_i \notin [\![\varphi_i]\!]$ or $o_i \notin [\![\varphi_j]\!]$. By Definition 10, $P(\varphi_i \vee \varphi_j) = b$ iff $\tau(\varphi_i \vee \varphi_j) = b$. By Definition 7, $\tau(\varphi_i \vee \varphi_j) = \frac{|[\![\varphi_i \vee \varphi_j]\!]|}{|O|}$. By Proposition 1, it follows that $[\![\varphi_i \vee \varphi_j]\!] = [\![\varphi_i]\!] \cup [\![\varphi_j]\!]$. This implies that $|[\![\varphi_i \vee \varphi_j]\!]| = |[\![\varphi_i]\!] \cup [\![\varphi_j]\!]|$. By the initial assumption, it follows that $[\![\varphi_i]\!] \cap [\![\varphi_j]\!] = \emptyset$. Thus, every $o_i \in [\![\varphi_i]\!] \cup [\![\varphi_j]\!]$ contributes only once to the cardinality of $|[\![\varphi_i]\!] \cup [\![\varphi_j]\!]|$, and coming from just one of the sets $[\![\varphi_i]\!]$ or $[\![\varphi_j]\!]$. This implies that $|[\![\varphi_i]\!] \cup [\![\varphi_j]\!]| = |[\![\varphi_i]\!]| + |[\![\varphi_j]\!]|$. By substitution, it follows that $\tau(\varphi_i \vee \varphi_j) = \frac{|[\![\varphi_i \vee \varphi_j]\!]|}{|O|} = \frac{|[\![\varphi_i]\!]| + |[\![\varphi_j]\!]|}{|O|}$. By splitting the fraction, it follows that $\tau(\varphi_i \vee \varphi_j) = \frac{|[\![\varphi_i]\!]|}{|O|} + \frac{|[\![\varphi_j]\!]|}{|O|}$. By Definition 10, it follows that $P(\varphi_i \vee \varphi_j) = P(\varphi_i) + P(\varphi_j)$. Since the model $\mathfrak{M}$ was chosen arbitrarily, this holds for every model.

Theorem 3 (Probability of Negation).

$$\models P(\neg\varphi) = 1 - P(\varphi)$$

Proof. Take an arbitrary model $\mathfrak{M}$ and an arbitrary outcome $o_i \in O \in \mathfrak{M}$. Take the tautology $\varphi \vee \neg\varphi$. Given the fact that it is a tautology, it follows that $\models (\varphi \vee \neg\varphi)$. By Proposition 4, it follows that $\models P(\varphi \vee \neg\varphi) = 1$. It also holds that $\models_{\mathfrak{M}} \neg(\varphi \vee \neg\varphi)$. By Theorem 2, it follows that $P(\varphi \vee \neg\varphi) = P(\varphi) + P(\neg\varphi)$. By substitution, it follows that $P(\varphi) + P(\neg\varphi) = 1$. By algebra, it follows that $P(\neg\varphi) = 1 - P(\varphi)$. Since the model and the outcome were chosen arbitrarily, this holds for any outcome and any model.

Theorem 4 (Probability of Equivalence).

$$If \models \varphi_i \leftrightarrow \varphi_j, then \models P(\varphi_i) = P(\varphi_j)$$

Proof. Take an arbitrary model $\mathfrak{M}$. Assume that $\models_{\mathfrak{M}} \varphi_i \leftrightarrow \varphi_j$. From the assumption, it follows that $[\![\varphi_i]\!] = [\![\varphi_j]\!]$ (if the formulas are satisfied in the same outcomes of the model, they must have the same validating set). Therefore, the conclusion follows directly by Definitions 7 and 10, since $\frac{|[\![\varphi_i]\!]|}{|O|} = \frac{|[\![\varphi_j]\!]|}{|O|}$ (the equivalence follows easily from the fact that the denominators are equivalent and so are the numerators of the fractions). The previous fact directly implies that $P(\varphi_i) = P(\varphi_j)$. Since the model was chosen arbitrarily, the same holds in every model.

Theorem 5 (Strong Additivity).

$$\models P(\varphi_i \vee \varphi_j) + P(\varphi_i \wedge \varphi_j) = P(\varphi_i) + P(\varphi_j)$$

Proof. Take the two tautologies $\varphi_i \leftrightarrow (\varphi_i \wedge \varphi_j) \vee (\varphi_i \wedge \neg\varphi_j)$ and $\varphi_j \leftrightarrow (\varphi_j \wedge \varphi_i) \vee (\varphi_j \wedge \neg\varphi_i)$. For simplicity, name $(\varphi_i \wedge \varphi_j)$ as α_1, $(\varphi_i \wedge \neg\varphi_j)$ as α_2, $(\varphi_j \wedge \varphi_i)$ as β_1 and $(\varphi_j \wedge \neg\varphi_i)$ as β_2.

Since the formulas taken are tautologies, it follows that $\models \varphi_i \leftrightarrow (\alpha_1 \vee \alpha_2)$ and that $\models \varphi_j \leftrightarrow (\beta_1 \vee \beta_2)$ By Theorem 4, we derive that $\models P(\varphi_i) = P(\alpha_1 \vee \alpha_2)$ and $\models P(\varphi_j) = P(\beta_1 \vee \beta_2)$. Note that $\models \neg(\alpha_1 \wedge \alpha_2)$ and that $\models \neg(\beta_1 \wedge \beta_2)$. By Theorem 2, it follows that $\models P(\alpha_1 \vee \alpha_2) = P(\alpha_1) + P(\alpha_2)$ and that $\models P(\beta_1 \vee \beta_2) = P(\beta_1) + P(\beta_2)$. By algebra and substitution of equivalents, it follows that $\models P(\varphi_i) + P(\varphi_j) = P(\alpha_1) + P(\alpha_2) + P(\beta_1) + P(\beta_2)$. Name this *fact one*.

Now, take the tautology $(\varphi_i \vee \varphi_j) \leftrightarrow (\alpha_1) \vee (\alpha_2) \vee (\beta_2)$. Since this is a tautology, it follows that $\models (\varphi_i \vee \varphi_j) \leftrightarrow (\alpha_1) \vee (\alpha_2) \vee (\beta_2)$. By Theorem 4, $\models P(\varphi_i \vee \varphi_j) = P(\alpha_1 \vee \alpha_2 \vee \beta_2)$. Also note that $\models \neg(\alpha_1 \wedge (\alpha_2 \vee \beta_2))$. By Theorem 2, this means that $\models P(\alpha_1 \vee \alpha_2 \vee \beta_2) = P(\alpha_1) + P(\alpha_2 \vee \beta_2)$. Note also that $\models \neg(\alpha_2 \wedge \beta_2)$. By Theorem 2, it follows that $\models P(\alpha_2 \vee \beta_2) = P(\alpha_2) + P(\beta_2)$. Therefore, $\models P(\alpha_1 \vee \alpha_2 \vee \beta_2) = P(\alpha_1) + P(\alpha_2) + P(\beta_2)$. By equivalence, this means that $\models P(\varphi_i \vee \varphi_j) = P(\alpha_1) + P(\alpha_2) + P(\beta_2)$ Name this *fact two*.

By putting together *fact one* and *fact two*, it follows that $\models P(\varphi_i \vee \varphi_j) = P(\varphi_i) + P(\varphi_j) - P(\beta_1)$, which, by algebra, is equivalent to $\models P(\varphi_i \vee \varphi_j) + P(\beta_1) = P(\varphi_i) + P(\varphi_j)$. By using the definition of β_1 and the commutativity of conjunction, it follows that $\models P(\varphi_i \vee \varphi_j) + P(\varphi_i \wedge \varphi_j) = P(\varphi_i) + P(\varphi_j)$.

4   Soundness and Completeness

We will now show that Λ is sound with respect to the class of all frequentist models.

4.1   Soundness

Theorem 6 (Soundness). *If* $\vdash_\Lambda \chi$ *then* $\models \chi$

Proof. In order to prove soundness, we will show that all axiom schemata are valid with respect to the class of all frequentist models and then show that MP preserves validity.

For instances of (i) and (ii) of Definition 12, the proof is straightforward and follows from the definition of propositional tautology and from the algebraic properties of the mathematical operations, which were assumed in frequentist models.

For instances of Ax. 1, note that the value of $P(\psi)$ is obtained by taking $\frac{\|\varphi\|}{|O|}$. By Definition 5, $\emptyset \subseteq [\![\varphi]\!] \subseteq O$. This implies that $|\emptyset| \leq |[\![\varphi]\!]| \leq |O|$. In addition, by Definition 3, it follows that $\emptyset \subset O$, which implies that $|\emptyset| < |O|$. Since $|\emptyset| = 0$, it follows that O is strictly positive. From those premises, it follows directly that

$\frac{\|[\varphi]\|}{|O|}$ is both defined ($|O|$ is different from zero) and is bigger or equal to zero, which proves the validity of Ax. 1.

For instances of Ax. 2, note that $[\![\top]\!] = O$. Since the two sets are equivalent, so is their cardinality, i.e., $\|[\![\top]\!]\| = |O|$. Moreover, the value of $P(\top)$ is equivalent to $\frac{\|[\![\top]\!]\|}{|O|}$. Given the fact that $|O|$ is finite and strictly bigger than 0, the fact that the numerator and denominator of $\frac{\|[\![\top]\!]\|}{|O|}$ are equivalent, implies that their ratio is equal to 1, which means that $P(\top) = 1$.

Instances of Ax. 3 are just instances of Theorem 2.

Instances of Ax. 4 are just instances of Theorem 4.

Finally, we must show that MP preserves validity. For the propositional formulas, this follows straightforwardly from propositional logic and the way the valuation function works. For probability formulas, it must be shown that if $P(\varphi_i) \odot b_m$ and $P(\varphi_i) \odot b_m \to P(\varphi_j) \odot b_t$, it follows that $P(\varphi_j) \odot b_t$.

Assume that $P(\varphi_i) \odot b_m$ and that $P(\varphi_i) \odot b_m \to P(\varphi_j) \odot b_t$. By Definition 10, the first assumption asserts that $\frac{\|[\varphi_i]\|}{|O|} \odot b_m$. By the definition of $\to$, the second assumption asserts that it must hold that either it is not the case that $\frac{\|[\varphi_i]\|}{|O|} \odot b_m$ or $\frac{\|[\varphi_j]\|}{|O|} \odot b_t$. By the first assumption, we know that it cannot be the first disjunct, therefore it must be the second one. However, the second one is just the definition of $P(\varphi_j) \odot b_t$. Thus, MP preserves validity.

4.2   Weak Completeness

To prove weak completeness, we will mimic the proof given in [7]. In order to do so, some facts must be established.

Since **FPL** is equivalent to propositional logic with the addition of probability formulas, and it is known that propositional logic is complete, we must only show that the added probability formulas also preserve completeness. This means, in particular, that we can focus only on the ψ formulas of the language. Thus, we must show that if a probability formula is valid, then it is provable.

Definition 14. (*Literals and Literal Formula*) We define as **literal** any atomic formula p_i or its negation $\neg p_i$. A **literal formula** δ is a formula $\varphi_1 \wedge \cdots \wedge \varphi_n$, where each φ_i is a literal.

Definition 15. We use the notation $At(\chi)$ to indicate the set of all atomic formulas contained in χ.

Lemma 1. *Let φ be a propositional formula. Call $Lit(\varphi)$ the set of all literal formulas δ obtained from $At(\varphi)$ such that $\delta \to \varphi$ is a propositional tautology. It is provable that $P(\varphi) = \sum_{\delta \in Lit(\varphi)} P(\delta)$.*

Proof. The proof will be given by induction on the size of $Lit(\varphi)$. Note that the main claim of Lemma 1 can be expanded as follows (where the Γ formulas are literals taken from $Lit(\varphi)$):

$$P(\varphi) = P(\varphi \wedge \Gamma_1) + \cdots + P(\varphi \wedge \Gamma_{2^i})$$

Base case. Assume that $i = 1$.

It must be shown that $P(\varphi) = P(\varphi \wedge \Gamma_1) + P(\varphi \wedge \Gamma_2)$ is provable. By the construction processes mentioned above, $\Gamma_2 := \neg \Gamma_1$ (Γ must be taken from the set of literal formulas, but since in the base of the induction, the set $Lit(\varphi)$ contains only one element, it is possible only to construct the two literal formulas Γ_1 and $\neg \Gamma_1$). Therefore, it must be proven that $P(\varphi) = P(\varphi \wedge \Gamma_1) + P(\varphi \wedge \neg \Gamma_1)$. This follows from Ax. 3 of Definition 12. Thus, the formula is provable.

Inductive base. Assume that the following is provable:

$$P(\varphi) = P(\varphi \wedge \Gamma_1) + \cdots + P(\varphi \wedge \Gamma_{2^i})$$

By Ax. 3, it is provable that $P(\varphi \wedge \Gamma_1) = P(\varphi \wedge \Gamma_1 \wedge p_{i+1}) + P(\varphi \wedge \Gamma_1 \wedge \neg p_{i+1})$. Using propositional reasoning (clause (i) in Definition 12) and instances of valid inequality formulas (clause (ii) in Definition 12), it is possible to substitute each formula $P(\varphi \wedge \Gamma_r)$ in the inductive base with a formula $P(\varphi \wedge \Gamma_r \wedge p_{i+1}) + P(\varphi \wedge \Gamma_r \wedge \neg p_{i+1})$. This implies that the inductive step is provable:

$$P(\varphi) = P(\varphi \wedge \Gamma_1) + \cdots + P(\varphi \wedge \Gamma_{2^{i+1}}).$$

As a particular case, the following formula follows:

$$P(\varphi) = P(\varphi \wedge \delta_1) + \cdots + P(\varphi \wedge \delta_{2^n}) \tag{2}$$

Now, take the set $At(\varphi)$, i.e., the set of all atomic formulas contained in φ. By propositional reasoning, if $\delta_r \in At(\varphi)$, then $(\varphi \wedge \delta_r) \leftrightarrow \delta_r$, which, by Ax. 4, means that $P(\varphi \wedge \delta_r) = P(\delta_r)$. If $\delta_r \notin At(\varphi)$, then $(\varphi \wedge \delta_r) = \bot$, which, by Ax. 4 and Theorem 1, means that $P(\varphi \wedge \delta_r) = 0$.

Given those facts, in Formula 2, each $P(\varphi \wedge \delta_r)$ can be either substituted with $P(\delta_r)$ or 0, depending on whether δ_r is included in $At(\varphi)$ or not. Therefore, Lemma 1 follows.

Theorem 7 (Weak Completeness). If $\models \chi$ then $\vdash_A \chi$

Proof. As previously stated, we can limit ourselves to probability formulas and their Boolean combinations, since the completeness of the portion of the language with propositional formulas follows from the completeness of propositional logic (note that our semantics does not differ from that of propositional logic). Moreover, due to Proposition 4, it is possible to reduce the problem of the provability of combinations of propositional and probability formulas to that of combinations of just probability formulas (which we will prove now).

We therefore prove completeness by showing that any consistent formula is satisfiable. For the proof, we will use χ to refer to Boolean combinations of probability formulas, thus excluding the portion of χ containing propositional formulas. A simple probability formula $P(\varphi) \odot b$ could always be read as a Boolean combination between $P(\varphi) \odot b \wedge P(\top) = 1$.

Assume that χ is consistent. First, transform χ into its disjunctive normal form (DNF), i.e., a disjunction of conjunctions (call each one of the disjunctive clauses χ_{clause_i}). Call this formula χ_{DNF} (χ_{DNF} will look like this: $\chi_{clause_1} \vee$

$\cdots \vee \chi_{clause_r}$). By propositional reasoning, it is possible to show that χ and χ_{DNF} are provably equivalent. Thus, since χ is consistent, so is χ_{DNF}. This implies that at least one of the clauses χ_{clause_i} must be consistent. Imagine that this is not the case, i.e., $\neg\chi_{clause_i}$ was provable for every i, then it is easy to see that $\neg(\chi_{clause_1} \vee \cdots \vee \chi_{clause_r})$ was also provable, which would make χ_{DNF} inconsistent, which is a contradiction. In addition to knowing that there is at least one χ_{clause_i} which is consistent, every model that satisfies χ_{clause_i} must also satisfy χ. We can therefore limit ourselves to the evaluation of a χ_{clause_i}. Call this formula f. Recall that all the elements of f are probability formulas ψ.

By Lemma 1, we know that we can construct a formula f' starting from f by substituting every conjunct in f, with a formula $P(\delta 1) + \cdots + P(\delta_{2^n})$, where $At(f)$ includes all the atomic propositions that appear in f and where the $\delta_1, \ldots, \delta_{2^n}$ are the $Lit(f)$.

Now, construct a formula f'' from f' by adding as conjuncts all the formulas $P(\delta_j) \geq 0$, with $1 \leq j \leq 2^n$, and the formula $P(\delta_1) + \cdots + P(\delta_{2^n}) = 1$. The first set of additions follows from Ax. 1, while the last addition is an instance of Lemma 1, where $\varphi = \top$, and Ax. 2.

This new formula f'' is provably equivalent to f' and therefore to f. To prove completeness, we therefore must show that f'' is satisfiable. Note that f'' is a conjunction of $2^n + r + s + 1$ formulas of the form:

$$P(\delta_1) + \cdots + P(\delta_{2^n}) = 1$$
$$P(\delta_1) \geq 0$$
$$\cdots$$
$$P(\delta_{2^n}) \geq 0 \tag{3}$$
$$P(\delta_1)_r + \cdots + P(\delta_{2^n})_r \geq b_r$$
$$P(\delta_1)_s + \cdots + P(\delta_{2^n})_s < b_s$$

where each $P(\delta_1)_r + \cdots + P(\delta_{2^n})_r \geq b_r$ captures the positive conjuncts of f', while each $P(\delta_1)_s + \cdots + P(\delta_{2^n})_s < b_s$ captures the negative conjuncts of f'.

Equation 3 can be easily transformed into a system of linear inequalities by substituting all instances of formulas with a variable (maintaining uniform substitution). Thus, the equation would become:

$$x_1 + \cdots + x_{2^n} = 1$$
$$x_1 \geq 0$$
$$\cdots$$
$$x_{2^n} \geq 0 \tag{4}$$
$$x_{1_r} + \ldots x_{2^n_r} \geq b_r$$
$$x_{1_s} + \ldots x_{2^n_s} \geq b_s$$

If Eq. 4 is satisfiable, so is f'', and, in turn, f', f and χ. Now, assume that f'' is unsatisfiable. This implies that Eq. 4 is unsatisfiable. Therefore, $\neg(f'')$ becomes an instance of (ii) from Definition 12. That would mean, however, that

f and therefore χ is inconsistent, which is a contradiction. Thus, f'' must be satisfiable.

Note that this proof shows, among other things, that the language specified in this paper (which resembles the one given in [7]) is not expressive enough to distinguish different models of probability. This follows from the fact that the language is complete both in terms of Frequentist Models and general probability models based on probability measures.

5 Conclusion and Future Works

In the paper, we presented a logic **FPL** to reason about probabilities with a relative frequency interpretation. We showed that it is possible to interpret the language of **FPL** with the standard semantics for propositional logic. **FPL** can give a peculiar frequentist interpretation of the probability operator as presented in [7]. We then gave a proof system for the language, proved that the traditional theorems of probability hold in our language, and established that the techniques employed in [7] to prove soundness and completeness work also for our interpretation.

In the future, we have two evolving plans. The first plan is to increase the expressiveness of **FPL** by evolving the logic in various directions. The second plan is to apply the logic to various contexts related to computer science and reasoning under uncertainty.

As far as the first plan goes, we would like: (i) to add a third truth value to the codomain of v, in order to express the irrelevance of a proposition to a given outcome. This would allow us to construct examples where only specific propositions (and not all of them) are tested during an experiment. (ii) to add multiple agents to the logic and to add communication channels between them. This would allow us to model scenarios in which different agents ran different experiments and then communicated their results to each other. (iii) to add dynamism to the language. This is the most interesting addition to the logic, since it would allow us to provide updating techniques that do not reduce to Bayesian updating. Specifically, we would like to construct a logic that allows a frequentist updating possibility. Obviously, also combinations of (i), (ii), and (iii) would be interesting additions to **FPL**.

As far as the second plan goes, we would like: (i) to employ **FPL** to reason about uncertainty in logics with trust operators, which could then measure the ratio of positive recommendations over all recommendations and of positive direct experiences (e.g., it would be interesting to evolve the propositional components of [2,3,15–17] to add a probabilistic part); (ii) to use **FPL** for the formal modeling of properties of real-world trust mechanisms based on quantitative notions of computational trust (e.g., [1,5]); (iii) to employ **FPL** to reason about probabilistic verification of machine learning, and statistical and approximate model checking.

A   Syntactic Proofs

Proof. *(Proof of theorem 1)*
Direct Proof.

1. $\neg(\top \wedge \bot)$ Taut.
2. $P(\top \vee \bot) = P(\top) + P(\bot)$ Ax. 3 from (1)
3. $P(\top) = 1$ Ax. 2
4. $(\top \vee \bot) \leftrightarrow \top$ Taut.
5. $P(\top \vee \bot) = P(\top)$ Ax. 4 from (4)
6. $P(\top \vee \bot) = 1$ Transitivity of $=$
7. $1 = 1 + P(\bot)$ Substitution from (2), (3), and (6)
8. $P(\bot) = 0$ By algebra over (7).

Proof. *(Proof of theorem 3)*
Direct Proof.

1. $P(\top) = 1$ Ax.2
2. $(\varphi \vee \neg\varphi) \leftrightarrow \top$ Taut.
3. $P(\varphi \vee \neg\varphi) = P(\top)$ Ax. 4 from (2)
4. $P(\varphi \vee \neg\varphi) = 1$ Substitution from (1) and (3)
5. $\neg(\varphi \wedge \neg\varphi)$ Taut.
6. $P(\varphi \vee \neg\varphi) = P(\varphi) + P(\neg\varphi)$ Ax. 3 from (5)
7. $P(\top) = P(\varphi) + P(\neg\varphi)$ Transitivity of $=$
8. $1 = P(\varphi) + P(\neg\varphi)$ Substitution over (1) and (7)
9. $P(\neg\varphi) = 1 - P(\varphi)$ By algebra over (8).

Proof. *(Proof of theorem 5)*
Direct Proof.

Note that the proofs of *Finite Additivity* and of *Equivalence of Probabilities* follow directly from the axioms of our proof system. You just have to assume the condition of the axioms and then get the conclusion directly by applying the relevant axiom.

1. $\varphi_i \leftrightarrow (\varphi_i \wedge \varphi_j) \vee (\varphi_i \wedge \neg\varphi_j)$ Taut.
2. $\varphi_j \leftrightarrow (\varphi_j \wedge \varphi_i) \vee (\varphi_j \wedge \neg\varphi_i)$ Taut.
3. $P(\varphi_i) = P((\varphi_i \wedge \varphi_j) \vee (\varphi_i \wedge \neg\varphi_j))$ Ax. 4 from (1)
4. $P(\varphi_j) = P((\varphi_j \wedge \varphi_i) \vee (\varphi_j \wedge \neg\varphi_i))$ Ax. 4 from (2)
5. $\neg((\varphi_i \wedge \varphi_j) \wedge (\varphi_i \wedge \neg\varphi_j))$ Taut.
6. $P((\varphi_i \wedge \varphi_j) \vee (\varphi_i \wedge \neg\varphi_j)) =$ Ax. 3 from (5)
    $= P(\varphi_i \wedge \varphi_j) + P(\varphi_i \wedge \neg\varphi_j)$
7. $\neg((\varphi_j \wedge \varphi_i) \wedge (\varphi_j \wedge \neg\varphi_i))$ Taut.
8. $P((\varphi_j \wedge \varphi_i) \vee (\varphi_j \wedge \neg\varphi_i)) =$ Ax. 3 from (7)
    $= P(\varphi_j \wedge \varphi_i) + P(\varphi_j \wedge \neg\varphi_i)$
9. $P(\varphi_i) + P(\varphi_j) = P(\varphi_i \wedge \varphi_j) + P(\varphi_i \wedge \neg\varphi_j)+$ Algebra and substitution
    $+P(\varphi_j \wedge \varphi_i) + P(\varphi_j \wedge \neg\varphi_i)$ from (3), (4), (6) and (8)
10. $(\varphi_i \vee \varphi_j) \leftrightarrow (\varphi_i \wedge \neg\varphi_j) \vee (\varphi_j \wedge \neg\varphi_i) \vee (\varphi_i \wedge \varphi_j)$ Taut.
11. $P(\varphi_i \vee \varphi_j) =$ Ax. 4 from (10)
    $= P((\varphi_i \wedge \neg\varphi_j) \vee (\varphi_j \wedge \neg\varphi_i) \vee (\varphi_i \wedge \varphi_j))$
12. $\neg((\varphi_i \wedge \varphi_j) \wedge ((\varphi_i \wedge \neg\varphi_j) \vee (\varphi_i \wedge \varphi_j)))$ Taut.
13. $P((\varphi_i \wedge \neg\varphi_j) \vee (\varphi_j \wedge \neg\varphi_i) \vee (\varphi_i \wedge \varphi_j)) =$ Ax. 3 from (12)
    $= P(\varphi_i \wedge \neg\varphi_j) + P((\varphi_j \wedge \neg\varphi_i) \vee (\varphi_i \wedge \varphi_j))$
14. $\neg((\varphi_j \wedge \neg\varphi_i) \wedge (\varphi_i \wedge \varphi_j))$ Taut.
15. $P((\varphi_j \wedge \neg\varphi_i) \vee (\varphi_i \wedge \varphi_j)) =$ Ax. 3 from (14)
    $= P(\varphi_j \wedge \neg\varphi_i) + P(\varphi_i \wedge \varphi_j)$
16. $P((\varphi_i \wedge \neg\varphi_j) \vee (\varphi_j \wedge \neg\varphi_i) \vee (\varphi_i \wedge \varphi_j)) =$ From (13) and (15)
    $= P(\varphi_i \wedge \neg\varphi_j) + P(\varphi_j \wedge \neg\varphi_i) + P(\varphi_i \wedge \varphi_j)$
17. $P(\varphi_i) + P(\varphi_j) = P(\varphi_i \vee \varphi_j) + P(\varphi_j \wedge \varphi_i)$ From (9), (11) and (16)

References

1. Aldini, A., Seigneur, J.M., Ballester Lafuente, C., Titi, X., Guislain, J.: Design and validation of a trust-based opportunity-enabled risk management system. Inf. Comput. Secur. **25**, 2–25 (2017)
2. Aldini, A., Curzi, G., Graziani, P., Tagliaferri, M.: Trust evidence logic. In: Vejnarová, J., Wilson, N. (eds.) Symbolic and Quantitative Approaches to Reasoning with Uncertainty: 16th European Conference (ECSQARU 2021). LNAI, vol. 12897, pp. 575–589. Springer (2021)
3. Aldini, A., Tagliaferri, M.: Logics to reason formally about trust computation and manipulation. In: Saracino, A., Mori, P. (eds.) Emerging Technologies for Authorization and Authentication. LNCS, vol. 11967, pp. 1–15. Springer (2020)
4. Antonelli, M., Lago, U.D., Pistone, P.: On counting propositional logic and wagner's hierarchy. In: Coen, C.S., Salvo, I. (eds.) Proceedings of the 22nd Italian Conference on Theoretical Computer Science. CEUR Workshop Proceedings, vol. 3072. Technical University of Aachen (2021)
5. Casadei, R., Aldini, A., Viroli, M.: Combining trust and aggregate computing. In: Cerone, A., Roveri, M. (eds.) Software Engineering and Formal Methods, SEFM 2017. LNCS, vol. 10729, p. 507–522. Springer (2018)

6. Demey, L., Kooi, B., Sack, J.: Logic and probability. In: Zalta, E.N., Nodelman, U. (eds.) The Stanford Encyclopedia of Philosophy. Metaphysics Research Lab, Stanford University, Fall 2023 edn. (2023)

7. Fagin, R., Halpern, J.Y., Megiddo, N.: A logic for reasoning about probabilities. Inf. Comput. **87**(1–2), 78–128 (1990)

8. Fattorosi-Barnaba, M., Caro, F.D.: Graded modalities. I. Stud. Logica. **44**(2), 197–221 (1985)

9. Fazlyab, M., Morari, M., Pappas, G.J.: Probabilistic verification and reachability analysis of neural networks via semidefinite programming. In: 2019 IEEE 58th Conference on Decision and Control (CDC), pp. 2726–2731 (2019)

10. Hérault, T., Lassaigne, R., Magniette, F., Peyronnet, S.: Approximate probabilistic model checking. In: Steffen, B., Levi, G. (eds.) Verification, Model Checking, and Abstract Interpretation, pp. 73–84. Springer (2004)

11. Hájek, A.: Interpretations of probability. In: Zalta, E.N. (ed.) The Stanford Encyclopedia of Philosophy. Metaphysics Research Lab, Stanford University, Fall 2019 edn. (2019)

12. Legay, A., Delahaye, B., Bensalem, S.: Statistical model checking: an overview. In: Barringer, H., Falcone, Y., Finkbeiner, B., Havelund, K., Lee, I., Pace, G., Roşu, G., Sokolsky, O., Tillmann, N. (eds.) Runtime Verification, pp. 122–135. Springer (2010)

13. O'Donoghue, B., Osband, I., Ionescu, C.: Making Sense of Reinforcement Learning and Probabilistic Inference (2020)

14. Pilipovsky, J., Sivaramakrishnan, V., Oishi, M., Tsiotras, P.: Probabilistic verification of ReLU neural networks via characteristic functions. In: Matni, N., Morari, M., Pappas, G.J. (eds.) Proceedings of The 5th Annual Learning for Dynamics and Control Conference, vol. 211, pp. 966–979. PMLR (2023)

15. Tagliaferri, M., Aldini, A.: From knowledge to trust: A logical framework for pre-trust computations. In: Gal-Oz, N., Lewis, P. (eds.) Trust Management XII. IFIP AICT, vol. 528, pp. 107–123. Springer (2018)

16. Tagliaferri, M., Aldini, A.: A trust logic for the varieties of trust. In: Camara, J., Steffen, M. (eds.) Software Engineering and Formal Methods. LNCS, vol. 12226, pp. 119–136. Springer (2020)

17. Tagliaferri, M., Aldini, A.: From belief to trust: a quantitative framework based on modal logic. J. Log. Comput. **32**(6), 1017–1047 (2022)

Using Maude to Model Motivation in Human Behaviour

Antonio Cerone$^{(\boxtimes)}$(iD) and Olzhas Zhalgendinov

Department of Computer Science, School of Engineering and Digital Sciences,
Nazarbayev University, Astana, Kazakhstan
antonio.cerone@nu.edu.kz, olzhas.zhalgendinov@nu.edu.kz

Abstract. Human beings act and think driven by motivation, which can be physiological as well as psychological. Although there is no unified theory of motivation, there are a number of theories in psychology that define conceptual models to explain distinct kinds and aspects of motivation. Such a conceptual fragmentation of the notion of motivation makes it very challenging the attempt to build a formal framework to model motivation. In this paper we address the needs underlying motivation, focusing in particular on physiological needs, such as the ones for food, water and sleep, and on the lowest level of psychological needs, the needs for safety and security. We use BRDL (Behaviour and Reasoning Description Language) to model human behaviour and thinking as well as its psychological motivation and LTSs (Labelled Transition Systems) to model physiological motivation. Finally, we illustrate our translation of this formal framework into the Maude rewrite language, which supports the simulation and analysis of the modelled cognitive systems.

Keywords: Behaviour and Reasoning Description Language (BRDL) · Labelled Transition Systems (LTSs) · Cognitive modelling · Rewriting logic · Maude · Theories of motivation

1 Introduction

Motivation is what causes us to act. It may be a rational reason, an intense desire, an emotional impulse, or a physiological need. Physiological needs, such as the ones for food, water and sleep, keep individuals alive. Desires and impulses are important aspects of the life of each individual and, in an evolutionist sense, they aim at the preservation of the species through social development, which foster collaboration and mutual support, and by intertwining with emotional states, such as love, which is essential for reproduction. Looking for rational reasons to act is a distinctive aspect of the human species and drives cultural and technological development.

Work funded by the School of Engineering and Digital Sciences (SEDS), Nazarbayev University, Astana, Kazakhstan.

This multifaceted nature of motivation makes it difficult to build a general theory of motivation. The first attempt, the *instinct theory* [13], aimed at identifying all possible instincts either physical (e.g., locomotion) or mental (e.g., curiosity). However, the number of proposed instincts became soon countless, thus making the theory too complicated. The focus moved then to the physiology of motivation, with the *drive theory* [11,12,18] emphasising the compelling urge (drive) to act in order to reduce physiological needs. But reducing needs is not enough to explain motivation, which made drive theory fall out of favour.

Explaining how to maintain the right balance between deficit and surplus is the objective of *homeostatic-regulation theory* [3]. This theory explains motivation as the tendency of the body to maintain a state of equilibrium (e.g., hunger is balanced by eating). Further theories try to address other aspect of motivation. *Opponent-process theory* [17] links motivation to emotion by explaining the acquisition of motivation as the result of a pattern of emotional experience (e.g., the motivation to use psychoactive drugs). According to *arousal theory* [2,19], the activity of the central nervous system determines the appropriate level of arousal for a given task in relation to the individual's personality (e.g., in general a low level of arousal would help in a complex task to prevent anxiety, but this is not the case for anxious personalities).

In this paper we aim at building executable models for human behaviour, which incorporate motivational aspects. We use a high-level notation, the Behaviour and Reasoning Description Language (BRDL) that allows psychologists and cognitive scientists to model and analyse human cognition in terms of their required attentional, reasoning and action components. With respect to our previous work on BRDL [6] and its use for modelling motivation generated by physiological needs [7], in this paper we consider the first two levels of Maslow's hierarchy of motivation [14,15], which include not only *physiological needs* but also *psychological* needs, such as safety and security.

BRDL has been implemented using the Maude rewrite language and toolset [16], thus providing a framework for the in silico simulation of human reasoning [9], some aspects of human learning [8,10] and the interaction with heterogeneous physical components [4,5]. However, motivation was not considered in such implementation. In this paper we extend the BRDL implementation by addressing the physiological aspects of motivation defined in our previous work [7] as well as psychological needs. Additionally, we also implement the use of variables in the BRDL framework, a feature needed to describe quantitative aspects of motivation, but which also greatly increases the framework expressiveness.

The rest of the paper is organised as follows. Sect. 2 defines our conceptual model for motivation and Sect. 2.1 illustrates it on the example of a user of a vending machine, which will follow us throughout the paper. Sect. 3 introduces BRDL syntax and Sect. 3.1 shows how to model motivation at a cognitive level. Sect. 4 revisits the environmental and physiological models introduced in our previous work [7], with reference to our current example. Sect. 5, after a short overview of Maude (Sect. 5.1), describes the implementation in Maude, including the translation of the BRDL syntax (Sect. 5.2), the infrastructure for modelling the variables (Sect. 5.3) and the rewrite rules that define the overall system behaviour (Sect. 5.4). Finally, Sect. 6 shows the results of analysing our example using Maude, draws conclusions and discusses possible future work.

2   A Conceptual Model of Motivation

According to the *hierarchical theory of motivation* proposed by Abraham Maslow [14,15] human needs can be organised into the following hierarchy:

1. *Physiological needs* are at the lowest level and are the basic, essential needs that allow individuals to live, such as the needs for food, water, sleep, which have as *physiological motivators* hunger, thirst, tiredness, respectively.
2. *Safety and security needs* aim at building and maintaining the appropriate living environment in which individuals are protected from environmental danger (safety) and social threats (security).
3. *Belongness and loving needs* aim at being accepted and cared by the society, that is, by the other individuals.
4. *Esteem needs* aim at feeling worthwhile, both in terms of self-appreciation and by comparing themselves to the other individuals.
5. *Self-actualisation needs* aim at fulfilling the human potential for purely hedonistic purposes, independently of external influences.

Level 1 is driven by physiological motivators, such as hunger, thirst and tiredness, while the higher levels are mostly driven by psychological motivators. Only when we have satisfied a specific level of needs, we move to the higher level. Thus, according to Maslow, we consider our safety and security only after having satisfied our physiological needs, such as food, water and sleep. In this paper we develop a model that covers the first two levels of Maslow's hierarchy.

In modelling physiological motivation, we adopt the *homeostatic-regulation theory* [3] which explains motivation as the tendency of the body to maintain a state of equilibrium (e.g., hunger is balanced by eating). However, this theory cannot explain higher-level needs, which are acquired through experience or exposure to a specific cultural environment. In this paper, we assume these needs already acquired.

2.1   Conceptual Model Example: The User of a Vending Machine

Let us consider an example that will follow us throughout the paper. Imagine being a user of a food vending machine in your office. You normally purchase a product from the machine every morning and you consume it later, when you become hungry. Thus, when you purchase the product, you are not driven by hunger, but by the will to be able to safely choose the product you prefer before it runs out and to avoid the lunchtime queue. Therefore, we can say that you have a level-2 need when you purchase and a level-1 need when you eat.

Suppose that one day you are trapped in an important meeting the whole morning and when you get out you are very hungry. You immediately run to the vending machine to purchase some food. This time you do not have a level-2 need when you purchase, but a level-1 need.

The vending machine is operated using a rechargeable card and a pin code, similarly to an automatic teller machine (ATM). You insert the card first and then you enter the pin code. If you have inserted the correct pin code, the card is returned first and, after you have collected it, the product is delivered. However, if you enter a wrong pin code, you get a warning and the option to either reenter the pin or abort the task. In this case, reentering a wrong pin within one hour will cause the card to be confiscated. If you abort, the card is returned and the product is not delivered. If you enter a wrong pin while you are not hungry, since you do not have a level-1 need, you can satisfy your level-2 needs and delay the purchase one hour in order to be safe from immediate confiscation, in case you entered a wrong pin again. But if you enter a wrong pin while you are hungry, then you are driven by a level-1 need and do not consider any level-2 needs. Thus, in this case, you reattempt the purchase with the risk of an immediate confiscation.

3   Formal Cognitive Model

BRDL models the content of *long-term memory (LTM)* in terms of *cognitive rules* (also called *LTM rules*) that either drive selective attention or represent *factual knowledge* or *procedural knowledge*. Cognitive rules drive the processing of information that has been transferred to *short-term memory (STM)* and may consist of facts retrieved from LTM, perceptions from the environment, action to be carried out on the environment and goals defining what you want to achieve. Thus STM acts as temporary store and is often called *working memory* (WM) when it is considered together with all its information processing functionalities.

Each cognitive rule has a general structure

$$g : info_1 \uparrow perc \Longrightarrow act \downarrow info_2$$

where

- g is a goal;
- $perc$ is a perception from the environment;
- act is an action performed on the environment;
- $info_1$ is the information to be removed from STM;
- $info_2$ is the information or goal to be stored in STM.

Symbol $\uparrow$ suggests removal from STM whereas symbol $\downarrow$ suggests storage in STM. We call *enabling* the part of the rule on the left of $\Longrightarrow$ and *performing* the part of the rule on the right of $\Longrightarrow$. The execution of a cognitive rule is enabled by the presence of goal g and information $info_1$ in STM, and by the perception $perc$ from the environment, and results in the removal of $info_1$ from STM, the performance of action act on the environment and the storage of new information $info_2$ in STM.

Information $info_1$ consists of a set of *basic items*, which are syntactically listed as sequence with a comma as a separator, but whose order is semantically irrelevant. Each basic item may be a perception, an action or a cognitive state. In addition to basic items, $info_2$ may also contain a goal. In fact, when the goal g is present in the rule, it is the only goal enabling the rule, while goals that may be in $info_2$ are actually produced in STM by performing the rule.

The syntax of goals is $goal(achievement)$, where $achievement$ is a set of basic items. The goal is achieved when $achievement$ contains the currently performed action or a basic item stored in STM. Once the goal is achieved, it is removed from STM. The absence of goal is denoted by $goal()$. In this case the syntax of a cognitive rule can be shortened as

$$info_1 \uparrow perc \implies act \downarrow info_2$$

When the goal is present and the perception is not, the control of attentional selection and behaviour is *deliberate* and is finalised to accomplish the information that is the arguments of the goal g. When the perception is present and the goal is not, the control of attentional selection and behaviour is *automatic*. When both the goal and the perception are not present, the rule models a mental inferential process consisting in the replacement of $info_1$ with $info_2$. When both the goal and the perception are present, the control of attentional selection and behaviour is *hybrid*, that is, driven by the goal (there is a deliberate decision to act or carry out a mental or attentional process), but reactive to the perception (the modality of acting or processing is automatic).

3.1   Cognitive Model Example: The User of a Vending Machine

The cognitive model of the vending machine user described in Sect. 2.1 is formalised in BRDL as follows:

$$goal(collect\ food, abort\ collect\ food) : \uparrow requested\ card$$
$$\implies insert\ card \downarrow expect\ requested\ pin \tag{1}$$
$$goal(eat) : \uparrow requested\ card \implies insert\ card \downarrow expect\ requested\ pin \tag{2}$$
$$expect\ requested\ pin \uparrow requested\ pin \implies enter\ pin \downarrow expect\ returned\ card \tag{3}$$
$$expect\ returned\ card \uparrow returned\ card$$
$$\implies collect\ card \downarrow expect\ delivered\ food \tag{4}$$
$$expect\ returned\ card \uparrow wrong\ pin\ warning$$
$$\implies \downarrow wrong\ pin\ warning \tag{5}$$
$$goal(collect\ food, abort\ collect\ food) : \ wrong\ pin\ warning \uparrow$$
$$\implies abort \downarrow abort\ collect\ food, try\ in\ one\ hour \tag{6}$$
$$goal(eat) : \ wrong\ pin\ warning \uparrow requested\ pin$$
$$\implies enter\ pin \downarrow expect\ returned\ card \tag{7}$$
$$abort\ collect\ food \uparrow returned\ card \implies collect\ card \downarrow \tag{8}$$
$$expect\ delivered\ food \uparrow delivered\ food$$

$$\longrightarrow collect\ food \downarrow food\ available \tag{9}$$
$$goal(eat):\ food\ available \uparrow \Longrightarrow eat \downarrow \tag{10}$$

The interaction is initiated by either rule 1 or rule 2. Rule 1 addresses level-2 needs by having the goal to either collect the food, to consume later, or abort the interaction, as a protection from card confiscation in case of wrong pin. Rule 2 addresses level-1 needs by having the goal to eat the food immediately. Both rules also require the perception that the vending machine requests the user to insert the card in order to be enabled and determine the action of inserting the card and the storage in STM of the expectation that the machine will request the pin next.

Rules 3 and 4 are driven by the matching expectations and perception that the machine requests the pin and returns the card, respectively, and determine the actions of entering the pin and collecting the card, respectively. Rules 5 models the implicit attention to the perception of the warning that a wrong pin has been entered, thus failing to meet the expectation.

Rules 6 and 7 model the reaction of the user to the warning. Rule 6 is driven by a goal expressing level-2 needs and aborts the interaction as a consequence of the warning thus avoiding card confiscation. Since the rule stores item *abort collect food* in STM and such an item is in the goal achievement, the goal is achieved and removed from STM. Then card collection occurs through rule 8, which does not have goal and models an automatic behaviour. Rule 6 also stores item *try in one hour* in STM, as a reminder for attempting the purchase again in one hour.

The goal removal caused by rule 6 prevents rule 1 from being enabled upon the request by the system to reenter the pin. However, a hungry user has goal $goal(eat)$, which expresses a level-1 need, thus enabling rule 7 once the system requests the user to reenter the pin. In this way, a user with level-1 ignores any level-2 needs.

Item *food available* is stored in STM by rule 9. Finally, rule 10 is enabled by the goal $goal(eat)$ and the food availability, and determines the action of eating the food. A user who is hungry while purchasing the food already has the goal $goal(eat)$ and eats the food immediately. Instead, a user who has purchased the food in advance will have the goal $goal(eat)$ at a later stage, when becoming hungry.

4   Formal Environmental Model and Physiological Model

With reference to the example in Sect. 3.1, we illustrate in Sect. 4.2 how the user interacts with the external environment and in Sect. 4.3 how goals (such as $goal(eat)$) are stored in STM as the result of physiological processes (such as becoming hungry). But first, in Sect. 4.1 we introduce *Labelled Transition Systems* (LTS) to model the *external environment* with which the user interacts as well as the *internal physiology* of the user.

4.1   Labelled Transition Systems(LTS)

We define an LTS by

- a set of perceptions;
- a set of invisible atomic states;
- an initial state consisting of a set of perceptions and a set of invisible atomic states;
- a set of transition rules $visible_1\ [invisible_1] \xrightarrow{act} visible_2\ [invisible_2]$, where sets of perceptions $visible_1$ and $visible_2$ and set of invisible atomic states $invisible_1$ and $invisible_2$ are represented by element separated by commas.

The system evolves starting from the initial state. Each transition rule models the transition from a source state consisting of a visible component $visible_1$ and an invisible component $invisible_1$ to a target state consisting of a visible component $visible_2$ and an invisible component $invisible_2$. The transition is triggered by action act.

4.2   External Environment: Interaction and Use of Variables

The interaction between the human component and the external environment is given through the synchronisation between a cognitive rule

$$g : info_1 \uparrow perc \implies act \downarrow info_2$$

and a transition rule

$$perc, visible_1\ [invisible_1] \xrightarrow{act} visible_2\ [invisible_2]$$

such that they share the same action act and the perception $perc$ of the cognitive rule is in the precondition of the transition.

The transition is enabled if the current state of the LTS includes $perc, visible_1$ as a subset of its visible component and $invisible_1$ as a subset of its invisible component. The transition changes the state of the LTS by replacing $perc, visible_1$ by $visible_2$ in its visible component and $invisible_1$ by $invisible_2$ in its invisible component. Note that $visible_2$ may contain $perc$.

For example, the interaction between the user modelled in Sect. 3.1 and a vending machine occurs through actions *insert card* (rules 1 and 2), *enter pin* (rule 3 and 7), *collect card* (rules 4 and 8), *abort* (rule 6) and *collect food* (rule 9). If we consider rule 4, the machine detection of the user's action of collecting the card is formalised by transition rule

$$returned\ card\ [\ ] \xrightarrow{collect\ card} delivered\ food\ [\ ] \tag{11}$$

Rule 4 and transition 11 share action *collect card*. The *returned card* perception of rule 4 is in the precondition of transition 11. If the machine is in state *returned card*, the transition, which has only visible part, is enabled and sychronises with rule 4, which has *returned card* as the perception. The transition execution on the shared action *collect card* changes the machine state to *delivered food* thus enabling rule 9.

As an example of use of variables in defining the environment, consider the situation in which the hungry user of our vending machine enters a wrong pin for the second time, thus causing a card confiscation. The LTS that models the vending machine must use a variable *pin attempts* to count the attempts to enter. Thus the transition that performs the card confiscation is modelled as follows:

$$requested\ pin\ [\ pin\ attempts\ = 1\ ] \xrightarrow{enter\ pin}$$
$$confiscated\ card\ [\ pin\ attempts\ + = 1\ ] \tag{12}$$

Note that on the left side of the transition variable *requested pin* is used within a condition, whereas on the right side of the transition it is assigned a value.

4.3 Interaction with the Internal Physiology

Interaction with the internal physiology is modelled through the direct effect of a transition rule on STM or through the transition being triggered by the content of STM. There are three kinds of transition rules that model the interaction with STM. Given information *info*, which may include goals,

- $visible_1\ [invisible_1] \xrightarrow{\downarrow info} visible_2\ [invisible_2]$ stores *info* in STM;
- $visible_1\ [invisible_1] \xrightarrow{info\uparrow} visible_2\ [invisible_2]$ removes *info* from STM;
- $visible_1\ [invisible_1] \xrightarrow{\downarrow info\uparrow} visible_2\ [invisible_2]$ is triggered by the presence of *info* in STM but does not change the content of STM.

As discussed in Sect. 2, we adopt the homeostatic-regulation theory [3], which explains motivation as the tendency of the body to maintain a state of equilibrium (e.g., hunger is balanced by eating). To this purpose, we identify the need with the motivator (the need of food is identified with its motivator 'hunger') and associate a numerical value with it. We consider two thresholds for the need, an activation threshold α and a saturation threshold σ such that $0 < \sigma < \alpha$.

We can suppose that initially the value of the need, which in our example is *hunger*, is below the α threshold. In this situation the motivator is inactive. The passing of time makes the need increase as a function of the human activity. When the need reaches the α threshold, the motivator becomes active. This means that we must carry out the appropriate activity, driven by a goal, to satisfy the need and, as a result, decrease its numerical value. Therefore, an iterative activity is carried out until the need has dropped down to the saturation threshold σ. Each step of the iterative cycle is driven by the goal and continues while the need is greater than the saturation threshold σ. In our example, the goal established in STM by an hungry person is 'eating' ($goal(eat)$) and it drives the deliberate behaviour of eating, which is modelled by cognitive rule 10 defined in Sect. 3.1. Once the need is as low as the saturation threshold σ, the motivator goes back to the inactive state.

Rule 10 models the cognitive aspects of hunger, that is, our deliberate eating activity. However, there are several physiological aspects that control the feeling of hunger and motivate us to eat and to stop eating. We use LTSs to model such physiological aspects.

With reference to our example, the *physiological motivation process* can be modelled using three transition rules:

activation $[hunger > \alpha, \ inactive] \longrightarrow [active]$
  This transition rule is enabled when condition $hunger > \alpha$ holds and the motivator state is *inactive*. The transition changes the state to *active*.

iteration $[hunger > \sigma, \ active] \xrightarrow{goal(eat)\downarrow} [active]$
  While condition $hunger > \sigma$, the motivator state is *active* and there is no goal $goal(eat)$ in STM, goal $goal(eat)$ keeps being stored in STM.

saturation $[0 \leq hunger \leq \sigma, \ active] \longrightarrow [inactive]$
  This transition rule is enabled when condition $0 \leq hunger \leq \sigma$ holds and the motivator state is *active*. The transition changes the state to *inactive*.

At the end of each iteration step of the physiological motivation process, the execution of cognitive rule 10 causes action eat to be performed. Since action eat is an argument of $goal(eat)$, the goal is achieved and thus removed from STM.

The *physiological satisfaction process* is determined by the feedback of the eating activity, which decreases the feeling of hunger. By denoting such a decrease by δ, we can model the satisfaction process as follows:

$$[hunger > \sigma] \xrightarrow{eat} [hunger - = \delta] \tag{13}$$

This transition rule is enabled when condition $hunger > \sigma$ holds and the transition occurrence decreases $hunger$ by a quantity δ.

We note that physiological states, such as needs, are modelled as invisible states since they are not directly visible from outside the LTS that models them. What is visible is the resultant behaviour, for example the fact that we eat.

5   Translation into Maude

5.1   A Short Overview of Maude

Maude [16] is a formal modelling language and high-performance simulation and model-checking tool for distributed systems. It makes use of

- algebraic equational specifications in a functional programming style to define data types;

– rewriting logic specifications, expressed using rewrite rules, to define the system evolution.

Maude *equational logic* supports declaration of *sorts*, with keyword **sort** for one sort, or **sorts** for many. A sort A may be specified as a subsort of a sort B by **subsort A < B**.

Operators are introduced with the **op** (for a single definition) and **ops** (for multiple definitions) keywords:

$$\text{op } f \: : \: s_1 \ldots s_n \text{ -> } s \: .$$
$$\text{ops } f_1 \: f_2 \: : \: s_1 \ldots s_n \text{ -> } s \: .$$

Operators can have user-defined syntax, with underbars '_' marking the argument positions. Some operators can have *equational attributes*, such as **assoc**, **comm**, and **id**, stating that the operator is associative, commutative and has a certain identity element, respectively. Such attributes are used by the Maude engine to match terms *modulo* the declared axioms. It is possible to declare the same operator on various subsorts (*subsort overloading*). In this case the **ditto** keyword may be used to specify the same equational attributes used in the previous declaration of that operator. Equational attributes **pred** and **gather** may be used to enforce precedence among operators. An operator can also be declared to be a *constructor* (**ctor**) that defines the carrier of a sort.

Axioms are introduced as equations using the **eq** keyword or, if they can be applied only under a certain condition, using the **ceq** keyword, with the condition introduced by the **if** keyword.

Variables used in equations are placeholders in a mathematical sense and cannot be assigned values. They must be declared with the keyword **var** for one variable, or **vars** for many. The use of the **owise** (or **otherwise**) equational attribute in an equation denotes that the axiom is used for all cases that are not matched by the previous equations.

Maude *rewrite rules*

$$\text{rl } [l] \: : \: t \text{ => } t' \quad \text{or} \quad \text{crl } [l] \: : \: t \text{ => } t' \text{ if } cond$$

define local transitions from state t to state t'. The second rule is a *conditional rewrite rules*, which is executed if condition *cond* is true.

Core Maude supports the definition of *functional modules*, which start with keyword **fmod** and end with keyword **endfm**, for algebraic equational specifications, and *system modules*, which start with keyword **mod** and end with keyword **endm**, for rewriting logic specifications. All Core Maude statements, apart from module definitions are ended by a dot. Modules can be imported using the keyword **protecting** followed by the name of the module ended by a dot.

Core Maude also enables module reusability with *parametrised modules*, which allow the use of such sorts as **List**, **Map**, and **Maybe**. These modules do not exist on their own, but are generated dynamically depending on the value domain for the elements of these sorts. The importation of the modules is performed by defining a view from the Core Maude's **TRIV** theory to a user-defined module and mapping sort `Elt` to a sort that defines the value domain for the elements of parametrized sorts.

One of the ways to check formal models in Maude is the `search` command: `search` t `=>*` t'. This command finds all terms that satisfy term pattern t' and can be reached by applying rewrite rules any arbitrary number of times starting from term t.

Full Maude is the object-oriented extension of Core Maude. It supports the definition of classes and objects within *object modules*, which start with keyword `omod` and end with keyword `endom`, and are enclosed between parentheses '(' and ')'. In fact, all commands and modules must be entered enclosed in parentheses when using Full Maude. A declaration `class` C | $att_1 : s_1, \ldots, att_n : s_n$ declares a *class* C with attributes att_1 to att_n of sorts s_1 to s_n. An *object* of class C is represented as a term $<O : C$ | $att_1 : val_1, ..., att_n : val_n >$ of sort `Object`, where O, of sort `Oid`, is the object's *identifier*, and where val_1 to val_n are the current values of the attributes att_1 to att_n.

5.2   BRDL and LTS Syntax and Manipulation Operators

The Maude Syntax for the basic items introduced in Sect. 3, the BRDL cognitive rules and the transitions that make up an LTS are defined in functional module `ENTITIES`, whose contents are described in this section.

Basic items are defined by sort `InfoItem`. They are the building blocks for cognitive rules, whose components are defined by the following sorts:

```
sorts InfoItem Goal Information ContentSTM ContentSTMList .
subsort String < InfoItem < Information .
subsorts InfoItem Goal < ContentSTM .
subsorts ContentSTM  Information < ContentSTMList .
```

Sort `InfoItem` includes strings (it has predefined sort `String` as a subsort), and models the basic items introduced in Sect. 3. In fact, we have been modeling items of information as strings (possibly of multiple words) throughout the paper.

We have seen in Sect. 3 that in the general form of cognitive rule

$$g : info_1 \uparrow perc \implies act \downarrow info_2$$

perc and *act* are basic items while $info_1$ and $info_2$ are sets of basic items, with $info_2$ possibly containing goals. Therefore, we define sorts `Information`, to model $info_1$ and `ContentSTMList`, to model $info_2$, as follows:

```
op noInfo : -> Information [ctor] .
op _,_ : Information Information ->
                  Information [ctor comm assoc id: noInfo prec 10] .
op _,_ : ContentSTMList ContentSTMList -> ContentSTMList [ditto] .
```

with `ContentSTMList` further characterised by the subsort declarations above stating that it has as subsort both `ContentSTM` (it may contain goals) and `Information` (it may contain basic items). A goal is defined as

```
op goalFromAnyOf : Information -> Goal [ctor] .
```

and the general form of cognitive rule is modelled as follows:

```
sort CognitiveRule .
op _:_|>_==_=>_>|_ :
        Goal Information InfoItem Parameter InfoItem ContentSTMList ->
                CognitiveRule [ctor] .
```

Similar definitions of constructors model the cases in which one of the components of the cognitive rule is missing, e.g.,

- `_:_|> ==_=>_>|_`, when the perception is missing;
- `_:_|>_==_=> >|_`, when the action is missing;
- `_:_|> ==_=> >|_`, when both the action and the perception are missing.

Object module COGNITION defines operators established and enabling to manipulate short-term memory (STM), whose content is modelled by the ContentSTMList sort, and sort LTM to model long-term memory (LTM) as a set of cognitive rules:

```
op established : Goal ContentSTMList -> Bool .
enabling : Information ContentSTMList -> Bool .
sort LTM .   subsort CognitiveRule < LTM .
op emptyLTM : -> LTM [ctor] .
op _;_ : LTM LTM -> LTM [ctor comm assoc id: emptyLTM)] .
```

Object module COGNITION also declares the class Cognition as follows

```
class Cognition | cognitiveLoad : Nat, stmCapacity : Nat,
                  shortTermMem : ContentSTMList, longTermMem :  LTM .
```

where attribute stmCapacity is the maximum number of basic items that can be in STM and cognitiveLoad models a number of additional items that are assumed to be in STM but are not modelled explicitly.

In order to model LTSs that can interact with the human cognition, we define a sort Event that characterises all events on which LTS and human cognition can synchronise and a sort SystemState that comprises a visible component consisting of the set of basic items and an invisible component consisting of a set of basic items 'hidden' within a sort Invisible:

```
sorts Event Invisible SystemState SystemRule .
subsort InfoItem < Event .
op _|> : ContentSTM -> Event [ctor prec 11 gather (e)] .
op |>_>| : ContentSTM -> Event [ctor prec 12] .
op >|_ : ContentSTM -> Event [ctor prec 11 gather (e)] .
op [_] : Information -> Invisible [ctor] .
op __ : Information Invisible -> SystemState [ctor] .
op __--_->__ : Information Invisible Event Information Invisible
                          -> SystemRule [ctor] .
```

Operators _|>, |>_>| and >|_ define the mutual effects between transition rules and STM, which are expressed in BRDL by *info* ↑, ↑ *info* ↓ and ↓ *info* respectively. Thus an event is either defined by one of these three operators or is an element of sort InfoItem, which is subsort of Event. The transition rules of the LTS are modelled by sort SystemRule. Their Maude syntax closely reflects the BRDL syntax presented in Sect. 3. The operator extSynch, which is defined as follows

```
op extSynch : InfoItem InfoItem -> Bool .
vars S1 S2 : String .
eq extSynch(S1,S2) = S1 == S2 .
```

checks whether two basic items are made up by identical strings. We show in Sect. 5.4 that this operator is used in rewrite rules to check whether the event S2 of a transition is the same as the action S1 in a cognitive rule thus enabling their synchronisation.

Object module LTS

```
(omod LTS is protecting ENTITIES .
    op emptyLTS : -> LTS [ctor] .
    op _;_ : LTS LTS -> LTS [ctor comm assoc id: emptyLTS] .
    sort LTS .    subsort SystemRule < LTS .
    class System | currentState : SystemState, transitions :   LTS .
endom)
```

defines the sort LTS as a set of transition rules by including SystemRule as a subsort and defining the ';' operator to construct the set. Class System defines the LTS as the current state, which will be initialised by the initial state when creating the object, and a set of transitions.

Finally, object module HUMAN

```
(omod HUMAN is protecting COGNITION + LTS + CONFIGURATION .
    class Human | cognition : Oid, physiology : Oid .
endom)
```

defines the human component as a combination of cognition and physiology.

5.3   Introducing Variables

Our previous BRDL translation into Maude [5,8–10] does not make use of variables to define cognition and environment. This results in models with duplicated cognitive rules and transitions, since each possible value of the same entity requires a dedicated rule/transition. This problem also involves physiological entities, such as *hunger*, whose value changes continuously controlled by the activation and saturation thresholds.

Therefore, we extend the standard BRDL logic to match and apply LTS transitions on information containing items with variable arguments. Variables allow the use of placeholders in a transition to generalise integer values in the current state of a system. This section describes how we map values from a system state to variables in the left-hand side of a transition and how we generate a new state from the right-hand side of the transition.

First, transitions with variables have a different matching logic. Therefore, we introduce the GENERIC-ENTITIES module, which extends the ENTITIES module to construct transitions from information containing variables. We declare a sort GenericItem as an extension of the InfoItem sort. GenericItem defines information items that may or may not contain variables. Similarly, the GenericInfo sort is a supersort of Information containing both GenericItem and InfoItem.

Then we define a logic for matching variables. The variables operate by replacing actual values in the system state. Therefore, we first introduce a functional module MATCHING to define basic concepts by using sorts Variable and Value to represent placeholders and actual contained values, respectively. These sorts are not populated in the module to support the use of any data types of values and any format of variables. Then, a new sort Matching is introduced to reflect the substitutions made when values are matched with variables:

```
sorts MatchingItem Matching Variable Value .
subsort MatchingItem < Matching .
op _:=_ : Variable Value -> MatchingItem [ctor prec 50] .
op noMatch : -> Matching [ctor] .
op _;_ : Matching Matching ->
                     Matching [ctor comm assoc id: noMatch prec 51] .
```

Finally, we define how the matching is constructed and how it is applied to the given system state, left-hand side and right-hand side of a transition to generate a new system state. Functional module GENERICS contains the definition of the genericMatch operator to generate matching for transition application:

```
protecting (MAYBE * (op maybe to failMatch)){Matching} .
op genericMatch : Matching Information GenericInfo  -> Maybe{Matching} .
```

This operator takes initial matching, current system information and the left-hand side of a transition as arguments. The resulting sort Maybe{Matching} means that the operator returns either a valid Matching of variables in a transition onto values in the system information or a term failMatch. This is defined by the importation of parametrised module MAYBE with mapping of the maybe operator onto failMatch. The operator is reduced according to the equation:

```
ceq genericMatch(MATCH, (INFOITEM, INFO), (ITEM, GENERIC))
= matchItem(MATCH, INFOITEM, ITEM);
  genericMatch((MATCH); matchItem(MATCH, INFOITEM, ITEM), INFO, GENERIC)
if matchItem(...) :: Matching /\ genericMatch(...) :: Matching .
```

The equation defines the information matching by applying matchItem on pairs of generic items ITEM on the left-hand side of a transition and information items INFOITEM in the system state. The condition ensures that this equation is applied only if matchItem of all pairs of items produces valid matchings. Otherwise, a different order of items will be considered because the information is a set and any item can match ITEM and INFOITEM.

The resolveGeneric operator transforms the system state depending on the generated matching and right-hand side of the transition:

```
crl [GENERIC_SYSTEM_EVOLUTION] :
  < TS : System | currentState : (INFO1, INFO3 [ INV1 , INV3 ]),
    transitions : (INFO2 [ INV2 ] -- auto -> INFO4 [ INV4 ]) ; TRANS >
=>
  < TS : System | currentState :
      (resolveState( INFO1, INFO2, INFO4 ), INFO3
        [ resolveState( INV1, INV2, INV4 ) , INV3 ]),
    transitions : (INFO2 [ INV2 ] -- auto -> INFO4 [ INV4 ]) ; TRANS >
if checkState( INFO1, INFO2, INFO4 ) /\ checkState( INV1, INV2, INV4 )
```

Fig. 1. Rewrite Rule for autonomous system evolution.

```
protecting (MAYBE * (op maybe to failResolve)){Information} .
op resolveGeneric : Matching GenericInfo -> Maybe{Information} .
```

As we noticed in Sect. 4.2, variables may be used within either conditions in the left-hand side of a transition or assignments in the right-hand side of a transition. This is implemented by the following sort and operator declarations:

```
sorts Condition ValueExpr GenericExpr .
subsort ValueExpr Variable < GenericExpr < GenericArgument .
subsort Value < ValueExpr < ExactArgument .
op _given_ : GenericExpr Condition -> GenericArgument .
op _'-_ : GenericExpr GenericExpr -> GenericExpr [prec 33 gather (E e)] .
op _'+_ : GenericExpr GenericExpr -> GenericExpr [prec 33] .
```

The operators '+ and '- contain a quote symbol to avoid conflicts with predefined operators for numbers, a common approach used in Maude [1].

As examples of use of variables, transition 12 introduced in Sect. 4.2 is translated to Maude code as

```
"requested pin" [ pinAttempts(a given a '== 1) ] -- "enter pin" ->
                        "confiscated card" [ pinAttempts(a '+ 1) ]
```

and transition 13 introduced in Sect. 4.3 is translated to Maude code as

```
noInfo [ hunger(a given a '> 0) ] -- "eat" -> noInfo [ hunger(a '- 1) ]
```

5.4   Rewrite Rules

The evolution of the Maude model is defined by the GENERIC-EVOLUTION system module, which consists of the following rewrite rules:

GENERIC_SYSTEM_EVOLUTION which is shown in Fig. 1, defines the autonomous evolution of a system modelled by an LTS that does not interact with a human component. The full definition of the System class of module LTS is given in Sect. 5.2. The rewrite rule checks whether in the LTS TS there is a transition

```
crl [GENERIC_PHYSIOLOGY_REMOVES_INFO] :
  < H : Human | cognition : CO, physiology : PHY >
  < CO : Cognition | cognitiveLoad : CL, stmCapacity : CAP,
    shortTermMem : (STMEM); CONTENTSTM1,
    longTermMem : LTMEM >
  < PHY : System | currentState : INFO1, INFO3 [ INV1 , INV3 ],
    transitions :
      (INFO2 [ INV2 ] -- (CONTENTSTM1 |>) -> INFO4 [ INV4 ]) ; TRANS >
=>
  < H : Human | cognition : CO, physiology : PHY >
  < CO : Cognition | cognitiveLoad : CL, stmCapacity : CAP,
    shortTermMem : (STMEM),
    longTermMem : LTMEM >
  < PHY : System | currentState :
    (resolveState( INFO1, INFO2, INFO4 ), INFO3
      [ resolveState( INV1, INV2, INV4 ) , INV3 ]),
    transitions :
      (INFO2 [ INV2 ] -- (CONTENTSTM1 |>) -> INFO4 [ INV4 ]) ; TRANS >
if CL + load(STMEM) <= CAP
   /\ checkState( INFO1, INFO2, INFO4 ) /\ checkState( INV1, INV2, INV4 )
```

Fig. 2. Rewrite Rule for a physiology removing information from STM.

whose source state INFO2 [INV2] matches the current state (with TRANS being the rest of the transitions) using the checkState operator and, if the matching is found, performs the transition using the resolveState operator to change the current state. Each of these two operators is applied separately to the invisible and visible parts of the current state and exploits the operators genericMatch and resolveGeneric discussed in Section 5.3 to respectively generate a valid Matching from mapping values INFO1 and INV1 onto generic INFO2 and INV1, respectively, and successfully resolve all variables in INFO4 and INV4.

GENERIC_PHYSIOLOGY_REMOVES_INFO which is shown in Fig. 2, is one of the three rewrite rules that define the interaction of cognition with physiology. This rewrite rule defines the effect of operator *info* |>, which implements *info* ↑. The definitions of Human, Cognition are given in Sect. 5.2. If item CONTENTSTM1 is in STM, then the rule is applied to a transition with event CONTENTSTM1 |> by removing CONTENTSTM1 from STM and performing the transition as in rewrite rule GENERIC_SYSTEM_EVOLUTION.

GENERIC_PHYSIOLOGY_ADDS_INFO defines the effect of operator >| *info*, which implements ↓ *info*, by adding item *info* to STM and performing the matching transition as in rewrite rule GENERIC_SYSTEM_EVOLUTION.

GENERIC_PHYSIOLOGY_READS_STM defines the effect of operator |> *info* >|, which implements ↑ *info* ↓, by checking whether item *info* is in STM without changing the contents of STM and performing the matching transition as in rewrite rule GENERIC_SYSTEM_EVOLUTION.

`GENERIC_EXTERNAL_SYNCHRONIZATION` defines the interaction between a human
   component and an external environment, such as user interfaces. It implements the synchronisation process described in Sect. 4.2.

`GENERIC_ACTION` defines the interaction between a human component and an
   external environment without any human perception of the state of the environment.

The last two rules use the operator `extSync` defined in Sect. 5.2 to check whether
the event in the transition of the external environment is the same as the action in
the cognitive rule of the human component thus enabling their synchronisation.

6   Conclusion and Future Work

We have extended our framework for modelling and analysing human reasoning
and behaviour [6,7] with a generalised notion of goal, which allows the choice
between alternative achievements. For example, this allows a user interacting
with an interface to consider the task completed either by succeeding or aborting.
This means to address level-2 psychological needs by safely aborting the task
when a danger or threat is perceived.

We also translated into Maude the formal framework for emotions and motivators defined in previous work [6,7] by incorporating it in the previous BRDL
translation [5,8–10] and further generalising it with the introduction of variables.

The BRDL translation into Maude defined in this paper has been used to
model and analyse the user task defined in Sect. 3.1. We used Maude model-
checking capabilities to verify that the absence of level-1 needs results in a
safer choice that meets level-2 needs. Starting from the initial cognitive state
$goal(collect\ food, abort\ collect\ food)$, which addresses level-2 needs, we can verify that the card is confiscated only when the user is hungry. We use Maude
`search` command to find a counterexample to this property. The searched term
models a vending machine state that contains the *confiscated card* basic item
and a physiology state that contains *inactive* hunger item. The result of the
search given by Maude is `No solution`, thus verifying the property. The Maude
code of the example can be downloaded from GitHub.[1]

Although it was not described in this paper for space reasons, the extension of
the Maude translation supports the generation of emotion defined using BRDL
and LTSs in our previous work [7]. In our future work we are planning to model
the effect of generated emotion on motivation and decision making.

References

1. Alpuente, M., Ballis, D., Romero, D.: A rewriting logic approach to the formal
   specification and verification of web applications. Sci. Comput. Program. **81**, 79–
   107 (2014)

---

[1] https://antoniocerone.github.io/Publications/2023/CIFMA.

2. Anderson, K.L.: Arousal and the inverted-uy hypothesis: Aq critique of nessiss's "reconceptualizing arousal". Psychol. Bull. **17**, 96–100 (1990)
3. Cannon, W.B.: The Wisdom of the Body. Norton (1932)
4. Cerone, A.: A cognitive framework based on rewriting logic for the analysis of interactive systems. In: Software Engineering and Formal Methods (SEFM 2016). Lecture Notes in Computer Science, pp. 287–303, No. 9763. Springer (2016)
5. Cerone, A.: Towards a cognitive architecture for the formal analysis of human behaviour and learning. In: STAF collocated Workshops 2018 (FMIS). Lecture Notes in Computer Science, pp. 216–232, No. 11176. Springer (2018)
6. Cerone, A.: Behaviour and reasoning description language (BRDL). In: SEFM 2019 Collocated Workshops (CIFMA). Lecture Notes in Computer Science, vol. 12226, pp. 137–153. Springer (2020)
7. Cerone, A.: A BRDL-based framework for motivators and emotions. In: SEFM 2023 Collocated Workshops (CIFMA). Lecture Notes in Computer Science, vol. 13765, pp. 351–365. Springer (2023)
8. Cerone, A., Murzagaliyeva, D.: Information retrieval from semantic memory: BRDL-based knowledge representation and Maude-based computer emulation. In: SEFM 2020 Collocated Workshops (CIFMA). Lecture Notes in Computer Science, vol. 12524, pp. 150–165. Springer (2021)
9. Cerone, A., Ölveczky, P.C.: Modelling human reasoning in practical behavioural contexts using Real-Time Maude. In: FM'19 Collocated Workshops-Part I (FMIS). Lecture Notes in Computer Science, vol. 12232, pp. 424–442. Springer (2020)
10. Cerone, A., Pluck, G.: A formal model for emulating the generation of human knowledge in semantic memory. In: Proceedings of DataMod 2020. Lecture Notes in Computer Science, vol. 12611, pp. 104–122. Springer (2021)
11. Hull, C.L.: Principles of Behaviour. Appleton-Century-Crofts (1943)
12. Hull, C.L.: A behaviour system: An introduction to behaviour theory concerning the individual organism. Yale University Press (1952)
13. James, W.: Psychology. Holt (1890)
14. Maslow, A.H.: A theory of human motivation. Psychol. Rev. **50**, 370–396 (1943)
15. Maslow, A.H.: Motivation and Personality, 2nd edn. Harper (1970)
16. Ölveczky, P.C.: Designing Reliable Distributed Systems. Undergraduate Topics in Computer Science, Springer (2017)
17. Solomon, R.L.: The opponent-process theory of motivation: The costs of pleasure and the benefits of pain. Am. Psychol. **35**, 681–712 (1980)
18. Woodworth, R.S.: Dynamic Psychology. Columbia University Press (1918)
19. Yerkes, R.M., Dodson, J.B.: The relation of strength of stimulus to rapidity of habit formation. J. Comp. Neurol. Psychol. **18**, 459–482 (1908)

Semantic Memory, Mnemonic Effort and Mnemonic Habit

Matthew Watts(✉) ⓘ

University of Miami, Coral Gables, FL 33416, USA
MXW685@miami.edu

Abstract. Semantic memory is often conceptualized as a storage space for an extensive assortment of explicit knowledge structures, formed as the result of a chronic, unconscious mechanism of abstraction and generalization. Against that, I argue that it is not the product of a dedicated system in which content is abstracted from experiences and stored, rather it's best understood as split between what can be called mnemonic efforts and mnemonic habits. Mnemonic efforts are the effortful expression of semantic knowledge and often take the form of the determination of similarities among multiple episodic memory traces activated in parallel. Mnemonic habits on the other hand form as a result of fluency of constructive memory processes that can be broadly understood as embodied, or enacted.

Keywords: Semantic Memory · Proceduralism · Memory

1 Introduction

Memory is involved in a bewildering variety of activities that differ on numerous dimensions, each with their own special conditions and characteristics and is presumed to underpin nearly all forms of human activity. In dealing with the complexity of memory and its uses, investigators have carved memory into various systems understood as serving two basic functions; that of remembering specific past experiences and the control of performance through the use of general knowledge and habituated action patterns [1, 2]. Of the many systems that have been proposed and the functions they serve, recent philosophical investigations of memory have disproportionately focused on episodic memory.

The relative scarcity of work on semantic memory in the philosophy of memory may be partially attributed to the influence of a pervasive yet seldom deeply analyzed view of semantic memory. This orthodox view maintains that semantic memory is a dedicated *storage space* of facts, which includes information such as the meanings and use of words, languages, and information about the world or a given situation and personal information. Furthermore, it's believed to be the result of a persistent, subconscious abstraction and generalization mechanism, possibly occurring during periods of inactivity or encoding [3].

Indeed, if the orthodox view were correct, there would be relatively little for philosophers to discuss regarding semantic memory. But acceptance of the orthodox view belies

A. Aldini (Ed.): SEFM 2023, LNCS 14568, pp. 90–98, 2024.
https://doi.org/10.1007/978-3-031-66021-4_6

fundamental issues with its conception. Despite ongoing research predicated on the existence of abstraction and generalization mechanisms, neuroanatomical structure, location, and memory storage systems, sustained issues surrounding the nature of abstraction and generalization mechanisms persist [4–7]. Moreover, the most widely accepted view of memory traces as distributed across multiple brain regions and sharing a network of neurons with several other traces has been argued to be in conflict with the concept of contentful memory traces [8–10], thus undermining the orthodox view of semantic memory as a storage space of explicit facts. Recent developments such as these in the philosophy of mind and cognitive science, as well as the apparent role of semantic memory in nearly all cognitive endeavors such as future oriented thought, counterfactual future thinking, and imagination, open an important space for an analysis of semantic memory.

The storage space model of semantic memory that presupposes unconsciously abstracted and generalized knowledge about the world attributable to the operation of a chronic abstraction mechanism taking place at encoding, or periods of inactivity may be a mistake. Instead I propose a non-storage based, contentless memory trace framework. I argue for a view of semantic memory that incorporates ideas from multi-trace memory frameworks, procedural knowledge representation in artificial intelligence and burgeoning enactive approaches. This approach provides a framework for semantic memory that integrates multiple bodies of research and provides us with new tools for tackling long standing conceptual issues of semantic memory and opens new avenues for discussion.

2   Semantic Memory and the Multiple Trace Approach

As early as (1923) Richard Semon used the metaphor of music homophony (a quality of music in which multiple different notes are being played surrounding the same melody) to describe a state when two or more memory traces are cued simultaneously [11]. In his characterizations Semon suggested two types of homophony that differ in their output; that of differentiating homophony and non-differentiating homophony [12]. Of interest here is the non-differentiating homophony which resulted when "similarities among engrams[1] are emphasized by mutual reinforcement of their common properties and mutual interference of their distinguishing ones." Semon described "non-differentiating homophony" as the more stable and common state and suggested that though multiple traces are nearly always retrieved, output from the memory system nonetheless could be in the form of separate memory traces, or some amalgamation of the separate memory traces [13].

While Semon never explicitly discussed homophony in regards to semantic memory, non-differentiating homophony provides us a characterization of abstract and/or general knowledge as constructed at the time of retrieval, as opposed to stored and retrieved. This idea was also at the heart of the multi-trace theories of memory that emerged during the 1980's as an alternative to the multi-system theories. For example, Hintzman developed the Minerva 2 model which demonstrated that abstract and general knowledge

---

[1] Semon uses the term 'engram' (indeed he originated the term) while there are subtle differences between the terms 'traces' and 'engram', they are generally used interchangeable in the literature. Semon [11, p. 248].

did not need to be stored, but rather could emerge through the functioning of a single system of storage [14]. This was further reinforced by experimental studies conducted by Whittlesea [1] showing that general and abstract knowledge could not be accounted for by general encoding routines thought to convert perception into knowledge [1]. General knowledge is instead constructed in response to the demands of particular tasks.

Older frameworks such as those proposed by Hintzman supposed that memory traces are specific to each experience, and therefore independent. This is problematic as, on the currently most widely accepted view, multiple traces are encoded with the same network of neurons. Traces then, are not individually locatable within the network of neurons and thus the causal path from experience to memory is blurred [10, 15]. But multi-trace models are not incompatible with the idea of traces sharing networks of neurons as recent multi-trace frameworks highlight the dynamic and blended nature of traces as a key feature. For instance, the Act-In framework highlights integration of traces, understood as a dynamic mechanism [16]. According to Versace et al. [16], trace activations occur continuously in parallel with the determination of similarities. Because traces share networks of neurons, each new trace modifies all traces in the network of neurons. This dynamical integration results in a constant reorganization of trace content through a mixture of environmental constraints, and activity of the rememberer.[2]

3   Proceduralism About Memory Causality

The process outlined above is in stark contrast to the often presupposed store of automatically, and unconsciously abstracted and generalized knowledge about the world attributable to the operation of a chronic abstraction mechanism taking place at encoding that characterizes most semantic memory theories. Multi-trace theory, though, only proposes a general operating principle of memory that can be implemented into various approaches to memory [15]. So, under a multi-trace theory of memory, it is unclear what the constructive processes operate on, or how they operate. On the multi-trace approach there is no content to abstract until trace activation, there are therefore no stored memory content. Semantic memory, then, is not retrieval of stored content, rather, it is construction of semantic knowledge.

Often, memory content is understood in componential terms, that is, in terms of the component parts that make up a memory, and ensure that it is experienced in a certain way [8]. For instance, as one remembers the camping trip to the mountains their memory will include various bits of information, such as "camping," "tent" and "campfire" that at least partly determine how the memory is experienced. On the componential view the causal link necessary for a memory to take place relies on a stored pattern of connections between nodes (corresponding to such things as concepts and event features) in a network of neurons. Recently it has been argued that the connections between nodes in the network amount to discrete packets of information carried from the initial experience creating the memory to the retrieved memory representation, and so, at least within distributed network accounts of memory, there is no way to track the causal history of memories

---

[2] Interestingly, the dynamical integration of the Act-In framework is equivalent to a continuous activation of reconsolidation taking place in everyday life. For more on the likelihood of this taking place see; Dudai (2004).

for particular past events, as each new memory involving those components will change the connection strengths within the network of neurons for each memory involving that component [10]. As Michaelian and Sant'Anna recently put it, "as far as the nature of memory traces is concerned, [combining a distributed conception of traces with the contentful conception of traces, is] incoherent" [8].[3]

Alternatively, we can account for the causality at play in remembering in terms of the procedural features displayed in the reconstruction of a memory. Perrin uses an analogy with a jigsaw puzzle to help illustrate:

> Let's imagine you have two copies of one and the same puzzle. The pieces are exactly the same in number, forms, and pictures in each box. Obviously, however alike the pieces are, as one makes one of the two puzzles one does not use the pieces of the other. In other terms, the two series of construction operations apply respectively to causally unrelated (though similar) bits of representation. But at the same time—this is a crucial point to my argument—these construction operations themselves can be causally related. For instance, if you make the two puzzles one after the other, you will possibly perform the second time better than the first time. And should you repeat the operations further, the enhancement will probably get ever clearer. So arguably, an earlier series of construction operations can get a later series enhanced, while the manipulated sets of pieces are distinct and causally unrelated [9].

On this view, causal relations are operative between the perceptual processes constructing the initial experience and the subsequent processes (re)constructing the memory of said experience. Just as causally connected operations can bring together causally unrelated sets of puzzle pieces into distinct copies of the same picture, causally connected constructive processes can bring about similar phenomenal experiences from causally unrelated bits of information. For instance, remembering pushing my son on the swing at Bryant Park displays distinct procedural features that reflect the previous experience in which I actually pushed my son on the swing at Bryant Park. What is retained on this view is not a stored representation of reality in the network of neurons, but rather dispositions to react through procedural abilities to construct representations [9]. This view receives theoretical and empirical support from a multi-trace framework outlined by Whittlesea [1] called SCAPE. On this framework, memory does not store information about what things are, instead it retains the way in which stimuli are operated on, and cognitive events are constructed. Memory should, then, be understood in procedural terms, in which it is the widely differing constructive processes (depending on such things as the broad diversity of contexts, availability of similar prior experiences, and current needs) that contribute to the multifarious kinds of knowledge typically attributed to multiple memory systems [1].

One might worry though that Perrin's puzzle analogy from above is inadequate to make this point, as it only repackages componential content as procedural skills. Afterall, the puzzle consists of component pieces the content of which is used to assemble the

---

[3] This is an important point, as while semantic memory is not stored content but rather the dynamic construction of knowledge, if Perrin is right, episodic memory is likewise not stored content.

overall puzzle. For example, imagine assembling a puzzle of a space shuttle lifting off. While various techniques will be used to assemble the puzzle correctly each relies on the content of the puzzle piece itself–think of matching up the right side of one piece containing the top of the rocket booster, with the left side of another piece containing the nose of the space shuttle, or matching up the tail of the shuttle with the blast beneath it. Or, using the shapes of the puzzle pieces themselves to construct the puzzle. In other words, Perrin's analogy implies that procedural causality develops the process through which individual packets of content are (re)constructed, and therefore merely adds an additional layer to the already problematic componential view.

Procedural causality, though, views causal relations as operative between the *processes* that constructed the initial experience and the *processes* that (re)construct the memory of the experience. Content on this view is not reassembled out of components like a puzzle, but rather created at the time of construction of the memory. Instead of puzzle pieces, then, we should analogize the procedural operations in question as the folding and sculpting techniques one uses in origami. Let's imagine that you set out to construct an origami crane from a blank sheet of paper. Performing the constructive operations on a sheet of paper to create the origami crane will enhance your ability to do so again the next time you construct a crane, and moreover, will enhance your ability to construct any origami figure involving those techniques. Importantly, unlike Perrin's previous puzzle piece analogy, the origami analogy allows us to envision causal relations that do not require the presupposition of the content of componential blocks implicated at each stage of construction. Rather, it is the constructive operations themselves (the folding and sculpting techniques) that create the content, and only after they have been completed are the various components of the crane–for instance the wing or the neck–produced.

4   Mnemonic Effort and Mnemonic Habit

As noted above, much expression of semantic memory seems to take place automatically and without conscious use. I've argued, though, that semantic memory is constructed in response to the demands of particular tasks and guided by the conscious use of prior experience and current thought. Consider, though, what procedural causality teaches us about causal relations in memory, namely, that they are operative between constructive *processes*, and not between *components*. Importantly, while focus thus far has been on the causal relations between the constructive processes of perception and the (re)constructive processes of memory these causal relations extend too *from* the (re)constructive processes of memory *to* the exploratory and constructive processes of perception. On this view, perception and memory result from the same types of activity and operate simultaneously. Storage and retrieval cannot then be considered as separate processes [16].

I thus propose that through the exploratory and constructive activities of perception and memory some of the behaviors we engage in during the constructive processes of perception become causally connected to our memory processes creating reusable embodied processes that enact knowledge through skillful operation of body schema.[4]

---

[4] The term used by researchers for this is "motor schema," I have opted for body schema to describe the phenomenon in terms in line with Merleau-Ponty. See; Halak (2018).

These embodied processes are habituated uses of mnemonic content and can be understood as on the one hand automatic certainties (or mnemonic habits–more on this below), and on the other hand as contributions to the construction of memory as a result of a coupling between the recovery situation and traces containing reusable processes activated on the basis of their similarity to the characteristics of the current circumstances of recovery. This view finds support from experimental studies highlighting the role of actions in the construction of memories[5] [17–22].

This indicates at least two distinct (but interrelated) ways of expressing semantic memory.[6] On the one hand there is knowledge expressed through habituated uses of mnemonic content; what we can call *mnemonic habits*. On the other hand there is the effortful expression of knowledge, the construction of which is guided by applicable prior experiences; what we can call *mnemonic efforts* [23].[7]

Much of our general knowledge arises from the fluent use of our procedural abilities that were once in service to mnemonic efforts. For example, were I prompted to give my name my response would take place without the conscious awareness that I know it. Likewise, my knowledge of what a bicycle looks like, and being able to correctly identify it from amongst various different transportation vehicles, often takes place without my need to attend to whether or not the object in front of me is a bicycle. Just as a young child may start out needing to put in effort to recall the next number, so too their ability to count will eventually become habitual–if enough time is spent with counting to habituate. Their familiarity with numbers will increase and thus their fluency. These are of course things that *can* be experienced as *known*, but that is not how we typically relate to them. Our relationship to them is one of automatic certainty. That is, this sort of semantic knowledge is one of habituated use, and our awareness of the knowledge is liminal. Perhaps at one point we needed to actively recall such things as our name, or how a bicycle looks, thus rendering them products of mnemonic effort, but in the course of everyday life, such facts cease to be the products of recall and instead reflect habituated uses of mnemonic content.

How we come to acquire these mnemonic habits can be analogized with Rowlands [24] proposed Rilkean memory, named after the poet Rainer Maria Rilke who described some memories as having become 'nameless' and 'forgotten' and changed into our 'blood' and 'glance and gesture'. Rilkean memories are the remnants of episodic memories that have lost their memory content but retained the act of remembering through acquired bodily dispositions, or affect. So the *act* of remembering has split from what is *remembered*, leaving Rilkean memory as an act of remembering without the original mental content. These bodily dispositions, behavioral affect, and sensations are

---

[5] See also (Christman et al. [17]; Brunye [18]; Brouillet et al. [19–21]; Brouillet et al. 2016; Camus [22]).

[6] While I discuss two ways of expressing semantic knowledge, I do not rule out the possibility of more ways in which it can be expressed. Moreover, This shares broad agreement with Moyal-Sharrocks (2009) distinction between habit/effort. The above differs in that I am discussing habits that develop from the use of semantic memory and conceptualized as fluency of the memorial content.

[7] In some cases the applicable prior episodes may amount to one or even multiple episodes synthesized to produce some general information.

thus constitutive of what remains of the original memory. As Rowlands puts it "A Rilkean memory arises when the act of remembering becomes divorced from what is remembered—because what is remembered has been lost" [24].

This approach to semantic memory is similar to procedural knowledge representation in artificial intelligence systems. Much like procedural representations, the procedural skills constituting mnemonic habits appear specialized to particular tasks, and highly efficient when performing said tasks. These procedural systems construct knowledge at the time of execution. Likewise, mnemonic habits aren't the explicit storage of general and abstract knowledge, instead they control performance by underpinning our interactions and exploration of the world.

Identifiable differences emerge, though, when conscious efforts to remember are considered. For example, many expressions of our semantic knowledge arise through the conscious use of prior experience to guide current thought and behavior. These are mnemonic efforts, the effortful expression of semantic knowledge. In some cases mnemonic effort takes the form of active attempts to recall through the use of such things as creative tactics or concentration [23]. Imagine, for instance, a young child attempting to remember their numbers. Often the child must actively work to recall the next number. Needing to run back through the string of numbers in order to remember that twelve comes after eleven. In other cases, mnemonic efforts involve the effortful use of prior experiences to guide current thought and behavior. In the case of mnemonic efforts what is achieved through engaging with the world is semantic knowledge, but the means by which it is attained requires a conscious use of prior experience to guide current thought and behavior.

As of now, procedural knowledge systems in artificial intelligence operate without an explicit sense of awareness. Their responses are driven by pre-defined rules and algorithms, and they lack the ability to reflect on past experiences or make subjective judgments and enact knowledge in the same way humans do. Environmental interaction, exploration, and embodiment play a crucial role in shaping human memory, and through this interplay, semantic knowledge emerges.

5   Conclusion

To sum up, as opposed to the standard conception of semantic memory as a storage space for knowledge structures, I instead proposed a proceduralist view of semantic memory that incorporates aspects of multi-trace memory frameworks, procedural knowledge representation in artificial intelligence, and burgeoning enactive memory approaches. Semantic memory is not a product of a dedicated system of storage, or the conscious awareness that accompanies the retrieval. On this view, we can understand memory as procedural knowledge driving performance in our interactions with the world. It is not the way that semantic memory is stored that determines its type, as semantic knowledge is not stored, instead it is the way in which it is constructed and subsequently used that determines its status as semantic memory.

Disclosure of Interests.   The authors have no competing interests to declare that are relevant to the content of this article.

References

1. Whittlesea, B.W.A.: Production, evaluation and preservation of experiences: Constructive processing in remembering and performance tasks. In: Medin, D.L. (ed.) The Psychology of Learning and Motivation, vol. 37, pp. 211–264. Academic Press, New York (1997)
2. Rowlands, M.: Memory. In: The Routledge Companion to Philosophy of Psychology, pp. 336–345. Taylor and Francis, New York (2009)
3. Dudai, Y., Karni, A., Born, J.: The consolidation and transformation of memory. Neuron **88**(1), 20–32 (2015)
4. Klein, S.: Making the case that episodic recollection is attributable to operations occurring at retrieval rather than to content stored in a dedicated subsystem of long-term memory. Front. Behav. Neurosci. **7**(3), 1–14 (2013)
5. Klein, S.: What memory is. WIREs Cogn. Sci. **6**(1), 1–38 (2015)
6. Duff, M.C., Covington, N.V., Hilverman, C., Cohen, N.J.: Semantic memory and the hippocampus: Revisiting, reaffirming, and extending the reach of their critical relationship. Front. Hum. Neurosci. **13**, 471 (2020)
7. Leboe-McGowan, J.P., Whittlesea, B.W.: Through the SCAPE looking glass: sources of performance and sources of attribution (2013)
8. Michaelian, K., Sant'Anna, A.: Memory without content? Radical enactivism and (post)causal theories of memory. Synthese 1–29 (2019)
9. Perrin, D.: A case for procedural causality in episodic memory. In: Michaelian, K., Debus, D., Perrin, D. (eds.) New Directions in the Philosophy of Memory, pp. 33–51. Routledge, New York (2018)
10. Robins, S.: Representing the past: memory traces and the causal theory of memory. Philos. Stud. **173**(11), 2993–3013 (2016)
11. Semon, R.W.: Mnemic Psychology. Macmillan, London (1923)
12. Schacter, D.L., Eich, J.E., Tulving, E.: Richard Semon's theory of memory. J. Verbal Learn. Verbal Behav. **17**(6), 721–743 (1978)
13. Hintzman, D.L.: Episodic versus semantic memory: a distinction whose time has come–and gone? Behav. Brain Sci. **7**(2), 240–241 (1984)
14. Yee, E., Jones, M.N., McRae, K.: Semantic memory. In: Stevens' Handbook of Experimental Psychology and Cognitive Neuroscience, vol. 3, pp. 1–38 (2018)
15. Versace, R., Labeye, E., Badard, G., Rose, M.: The contents of long-term memory and the emergence of knowledge. Eur. J. Cogn. Psychol. **21**(4), 522–560 (2009)
16. Versace, R., Vallet, G.T., Riou, B., Lesourd, M., Labeye, E., Brunel, L.: Act-In: an integrated view of memory mechanisms. J. Cogn. Psychol. **26**(3), 280–306 (2014)
17. Christman, S.D., Garvey, K.J., Propper, R.E., Phaneuf, K.A.: Bilateral eye movements enhance the retrieval of episodic memories. Neuropsychology **17**(2), 221–229 (2003)
18. Brunyé, T.T., Mahoney, C.R., Augustyn, J.S., Taylor, H.A.: Horizontal saccadic eye movements enhance the retrieval of landmark shape and location information. Brain Cogn. **70**(3), 279–288 (2009)
19. Brouillet, D., et al.: Sensory–motor properties of past actions bias memory in a recognition task. Psychol. Res. **79**(4), 678–686 (2015)
20. Brouillet, D., Milhau, A., Brouillet, T., Servajean, P.: Effect of an unrelated fluent action on word recognition: A case of motor discrepancy. Psychon. Bull. Rev. **24**(3), 894–900 (2017)
21. Brouillet, D.: Enactive Memory. Front. Psychol. **11** (2020)
22. Camus, T., Hommel, B., Brunel, L., Brouillet, T.: From anticipation to integration: the role of integrated action-effects in building sensorimotor contingencies. Psychon. Bull. Rev. **25**, 1059–1065 (2018)

23. Moyal-Sharrock, D.: Wittgenstein and the Memory Debate. New Ideas in Psychology Special Issue: Mind, Meaning and Language: Wittgenstein's Relevance for Psychology, vol. 27, pp. 213–227 (2009)
24. Rowlands, M.: Memory and the Self: Phenomenology, Science and Autobiography. Oxford University Press, United States of America (2017)

Scientific Understanding and the Explanatory Integration in Cognitive Sciences

Giovanni Galli[✉][iD]

University of Urbino, Urbino, Italy
g.galli13@campus.uniurb.it

Abstract. Scientific understanding in the field of cognitive sciences is a multifaceted concept that necessitates reflecting on the integration of various explanations. In this paper, I argue that different kinds of explanations regarding cognitive sciences can be integrated into an account of explanatory scientific understanding, as proposed by Khalifa. Moreover, I propose that scientific understanding should be distinct from mere knowledge and should be conceptualized as a nexus of explanation. This paper explores the theoretical foundations of scientific understanding, discusses different types of explanations in cognitive sciences, criticises a reduction problem in Khalifa's account and elucidates how these explanations can be effectively integrated to foster a holistic understanding of cognitive phenomena. Through an interdisciplinary approach, the aim of this paper is to enrich our comprehension of cognitive sciences and promote a more unified perspective on scientific understanding.

Keywords: Explanatory integration · Scientific understanding · Cognitive sciences

1 Introduction

Scientific understanding in the field of cognitive sciences is a multifaceted concept that necessitates reflecting on the integration of various explanations. In this paper, I argue that different kinds of explanations regarding cognitive sciences can be integrated into an account of explanatory scientific understanding, as proposed by Khalifa. Moreover, I propose that scientific understanding should be distinct from mere knowledge and should be conceptualized as a nexus of explanation. This paper explores the theoretical foundations of scientific understanding, discusses different types of explanations in cognitive sciences, criticises a reduction problem in Khalifa's account and elucidates how these explanations can be effectively integrated to foster a holistic understanding of cognitive phenomena. Through an interdisciplinary approach, the aim of this paper is to enrich our comprehension of cognitive sciences and promote a more unified perspective on scientific understanding.

A. Aldini (Ed.): SEFM 2023, LNCS 14568, pp. 99–113, 2024.
https://doi.org/10.1007/978-3-031-66021-4_7

Scientific understanding is a pivotal concept in the scientific domain and the specific area of cognitive sciences in the last decades has tremendously increased our understanding of cognitive phenomena. Since cognitive scientists aim to give an account of our cognitive life, namely how our minds work, how is it possible to understand a language, and even how it is possible to understand why the sky is blue or why decarbonization is urgent to slow down the climate change, philosophers engaged in the study of understanding and scientific understanding should be interested in focusing on their theoretical frameworks and methodologies. This is indeed the case of Khalifa et al. (2022), which advance a way to illuminate the explanatory integration issue in the Cognitive sciences on the base of Khalifa's account of scientific understanding. Given the interdisciplinary origins of Cognitive sciences, we find between their theoretical posits and methodological tools. In particular, cognitive sciences to comprehend the intricate workings of the human mind must employ diverse explanations that span various disciplines, such as psychology, neuroscience, philosophy, and artificial intelligence. Tracking back to the development of the cognitive sciences reveals two main approaches towards the object of research and the methodologies implied: unification and pluralism. These two approaches reflect the philosophical view about the unity of science as debated in the Vienna Circle. According to Neurath (1937), a defender of pluralism, sciences should have been coordinated. Carnap instead argues that all sciences should be reduced to one grand unifying theory. According to Gentner (2019), at the foundation of cognitive sciences, researchers were predominantly pluralistic, while views of reductive unity were prominent in the 1960s thanks to the manifesto of Oppenheim and Putnam (1958), but the received view on reduction cannot be applied to cognitive theories. In the last years, some scholars have argued that it is possible to attain a unified science of cognition "by showing how functional analyses of cognitive capacities can be and in some cases have been integrated with the multilevel mechanistic explanations of neural systems" (Piccinini and Craver 2011). The crucial problem remains that we do not have an efficient account of what explanatory integration entails (Miłkowski 2013, 2016). The canonical framework of cognitive science (singular) between the 70s and the 80s was given by computational functionalism. The first cognitive scientist to work on a solution to the issue of different explanations and the relations among them, he labelled it as the "level" of explanations, was Marr. His theory of vision is a paradigmatic case of cognitive science. He distinguishes three explanatory levels. At the higher level, it is defined as a high cognitive task as a function mapping some input over some output (this is the concept of computation defined as abstracting from the constituent operations). At the second level, the algorithm that computes the function is specified. The third level concerns what is called the neural hardware implementing the algorithm. This schema, albeit quite simple compared to the problem complexity, reveals the meaningful feature of classical cognitive science: its way of abstracting from the brain. Marconi (2001) adds another level of explanation to Marr's vision theory: the program level. The notion of "level" will be later crucial; it is Craver (2008) who develops this notion and argues that the theories of psychobiological sciences entail structures and dynamics in which the levels of mechanisms are fundamental. The explanatory levels are related in a hierarchical order between mechanisms explained from the higher to the lower level, i.e. we have the level of the spatial navigation function, then the hippocampus mechanism generating spatial maps, neurons inducing long term potentiation and finally, the activation of NMDA receptor (Craver 2005). From one unique framework (the functionalist in Marr's theory) to the mechanistic framework of Craver's claim, the common trait is that a unificatory purpose holds in the first paradigm of cognitive science. A consequence of this way of dealing with cognition is that, even if they distinguish different levels of explanations and their relations, no different kinds of explanations are involved and consequently, there is no need for integration. But in recent years, it became clear that the plurality of sciences involved in the study of cognition is so broad and structured and the cognitive phenomena are so complex that the post-classical studies, through the vertical neuroscientific expansion towards the brain and the horizontal one towards the body and the environment, cannot be subsumed under the singular term of science, but it is a family of sciences which collaboratively work on cognition, whit different theoretical backgrounds and methodological assumptions and instruments. In this panorama, it is clear that different sciences producing different kinds of explanations present the issue of how to integrate the explanations in order to assure a common scientific understanding (as a new basis for establishing a unified field of knowledge exchange, comparison and consensus) of the cognitive phenomena. Here, the scholars of understanding can come to help.

Khalifa's account of explanatory scientific understanding provides a framework to integrate these diverse explanations into a unified understanding. However, it is imperative to distinguish scientific understanding from mere knowledge and emphasize its role as a nexus of explanation. In this paper, I delve into the theoretical underpinnings of scientific understanding, elucidate the different types of explanations in cognitive sciences, and argue for the integration of these explanations to facilitate an integrative perspective on cognitive phenomena.

2   Contrasting Khalifa's and De Regt's Accounts of Scientific Understanding

Scientific understanding (SU) is a multifaceted concept that has garnered significant attention among philosophers of science. Two prominent scholars, Khalifa and De Regt, have proposed distinct accounts of scientific understanding that illuminate different aspects of this complex phenomenon. In this discussion, we will explore and contrast Khalifa's (2012, 2013a, 2013b, 2017, 2019, 2023a) account of scientific understanding with De Regt's (2005; 2017) perspective, shedding light on their fundamental differences.

Khalifa's account of scientific understanding emphasizes the importance of integrating diverse explanations to achieve a holistic grasp of natural phenomena. According to Khalifa, scientific understanding goes beyond mere knowledge acquisition; it involves gaining insight into the causal-mechanical, structural, and functional aspects of a phenomenon. This perspective contends that understanding arises when we appreciate the interplay of these three explanatory components. Khalifa's framework is rooted in the idea that scientific understanding is not a mere collection of facts but a deeper comprehension of the underlying mechanisms, organization, and purpose behind natural phenomena. Khalifa's account places a strong emphasis on interdisciplinary collaboration and the integration of different types of explanations from various scientific disciplines. It underscores the need to connect the dots between causal-mechanical, structural, and functional explanations, recognizing that a comprehensive understanding emerges when these facets are interwoven. It is one of the most demanding accounts of SU and it requires that experts evaluate their explanations according to the best available methods and evidence. There are three main principles supporting his model of SU, called Explanation, Knowledge and Science (EKS). Let's recall the main three principles he advances. The first is the Explanatory Floor (EF):

EF: Understanding why Y requires possession of a correct explanation of why Y.

Since scientific understanding can be subject to improvements, given its dynamical nature, Khalifa advances the Nexus Principle (NP):

NP: Understanding why Y improves in proportion to the amount of correct explanatory information about Y (=Y's explanatory nexus) in one's possession.

The third principle is the Scientific Knowledge Principle (SKP), which bounds the notion of scientific understanding to knowledge, namely scientific knowledge:

SKP: Understanding why Y improves as one's possession of explanatory information about Y bears greater resemblance to scientific knowledge of Y's explanatory nexus.

This last principle gives the idea that even the same explanatory information could be linked to different degrees of understanding, given the abilities and the information of relevant theories, models, empirical observation and experience scientists can have. Khalifa also gives a detailed definition of scientific knowledge of explanation (SKE):

SKE: An agent S has a scientific knowledge of why Y if and only if there is some X such that S's belief that X explains Y is the safe result of S's scientific explanatory evaluation (SEEing).

He concludes that thanks to SEEing and safety, the epistemological concept that requires an agent's belief to not easily have been given the way in which it was formed (Pritchard 2009), scientific knowledge of an explanation is achieved when "one's commitment to an explanation could not easily have been false given the way that one considered and compared that explanation to plausible alternative explanations of the same phenomenon" (Khalifa et al. 2022: 8). Khalifa and colleagues then argue that his framework of SU provides a "fruitful account how different explanations, such as the ones discussed above, can be integrated. The Nexus Principle is the key engine of integration" (Khalifa et al. 2022: 8).

In contrast, De Regt's contextualism cannot fit well to address the explanatory integration in Cognitive sciences. In fact, due to his Criterion of Understanding Phenomenon (CUP), supporting SU, he cannot account for the integration of different kinds of explanations. According to CUP: A phenomenon p is understood scientifically if and only if there is an explanation of p that is based on an intelligible theory T and conforms to the basic epistemic values of empirical adequacy and internal consistency. Given the multiplicity of the entangled kinds of explanations in cognitive sciences, we should eschew De Regt's contextual theory, if we want to improve the explanatory integration in cognitive sciences. De Regt's account of scientific understanding focuses also on the role of scientific models in the acquisition of understanding. He argues that understanding arises from an intimate familiarity with and a deep engagement in the use of scientific models. De Regt contends that scientific understanding is closely linked to our ability to manipulate, apply, and navigate these models effectively. De Regt's account places less emphasis on the integration of different types of explanations and more on the centrality of models in scientific practice. For De Regt, understanding is intimately tied to our ability to make predictions, explain phenomena, and solve problems using these models. The more proficient one can employ a model to achieve these goals, the deeper one understands the relevant scientific domain.

The critical distinction between Khalifa's and De Regt's accounts lies in their conceptions of what constitutes scientific understanding. Khalifa's perspective emphasizes a broader understanding that integrates various types of explanations, fostering a comprehensive grasp of complex phenomena. In contrast, De Regt's account narrows the focus to the practical utility of scientific models in achieving understanding. Additionally, Khalifa's account highlights the importance of interdisciplinary collaboration, encouraging the integration of insights from different scientific disciplines to enrich understanding. While acknowledging the importance of models, De Regt's account does not explicitly emphasize interdisciplinary integration to the same extent.

In summary, Khalifa and De Regt offer distinct perspectives on scientific understanding, with Khalifa focusing on the integration of diverse explanations and De Regt highlighting the role of scientific models in achieving understanding. These accounts provide valuable insights into the multifaceted nature of scientific understanding and offer different lenses through which we can explore and appreciate the richness of this concept in the philosophy of science.

3   Kinds of Explanations in Cognitive Sciences

Cognitive sciences are a plural area of inquiry specialized in human and artificial cognition. To the extent of this paper, the plural term "sciences" is chosen over the singular "cognitive science" to focus on the different scientific disciplines and methodologies involved in understanding the brain's functioning and the ability to produce the mind.[1] Since each scientific disciplines have its own methods and idiolects, they produce distinctive kinds of explanations, even if they aim

---

[1] See Marconi (2001) and Legrenzi (2002) for a discussion about the differences between the use of the singular and plural terminology.

to answer common questions about the brain and the mind, namely they all, as cognitive sciences, aspire to understand cognition. Each branch of cognitive sciences produces peculiar explanatory information, which shall be integrated to achieve scientific understanding of cognitive processes.

We can trace different kinds of explanations as an output of psychological, neuroscientific, linguistic, and computational research, each of them capturing information about dependency relations within the target system. Even if, since the early foundation of cognitive sciences, it was clear that the convergence of different research programs and their explanatory tool-box, had the same objective (Miller 1979), not all the researchers are convinced about the integrability of different explanatory information. Those defending the integration of mechanistic explanations "will take integrations of these explanations to be able to explain cognitive competencies such as language production and comprehension, memory, perception, problem solving, categorisation, and reasoning; but also general, flexible behaviours and real-time performance, as well as the processes of learning and development that are characteristic of the human cognitive system" (Taylor 2019: 4575–76). One of the main issues concerning using different kinds of explanation in the multifaced area of cognitive sciences is that the struggle of finding a criterium to distinguish between what is an explanation and what is not is highly demanding. The possibility of having different explanations, explanantia, for the same phenomenon E, the explanandum, poses a problem for the unificationists who are in search of a unifying criterium for explanation. On one hand, the unificationists wish to exclude some kinds of explanations from the role of explanans, so that they could find the explanation satisfying the criterium. On the other hand, pluralists allow to have many kinds of explanations, each one playing the role of explanans, offering different angles on the phenomenon E, which we can understand completely only if we integrate all the explanations at our disposal together. The latter rules out the unicity of the explanatory criterium the formers want to be satisfied by a single dominant explanation. Khalifa et al. (2022) explore the possibility of integration between mechanistic, computational, topological and dynamical explanations.

3.1 Mechanistic

Mechanistic explanations are the focus of the research of Bechtel and Richardson (1993), Machamer et al. (2000), Craver (2007), Illari and Williamson (2010), Glennan (2017), Craver and Tabery (2019). This kind of explanations is widespread in the cognitive sciences. There is no consensus on the proper characterization of mechanisms or how exactly they figure in mechanistic explanations, but one of the most used definitions comes from Glennan, which conceives mechanism as follows:

Glennan's definition of mechanism: A mechanism for a phenomenon consists of entities (or parts) whose activities and interactions are organized so as to be responsible for the phenomenon.

There are many cases in which this kind of mechanism is used in cognitive sciences, and one non-exhaustive example is the explanation we can found of action potential in neurons, also known as "nerve impulse" or "spike". Action potential occurs when there is a rapid rise and fall of the membrane potential of a certain cell. In particular, when the action potential travels down an axon, we can notice that the electric polarity across the membrane of the axon cells changes. In response to a signal from another neuron, sodium and potassium-gated ion channels open and close rapidly as the membrane hit its threshold potential. There is then depolarization and repolarization as the ion channels move into and out of the axon, creating a change in electric polarity between the outside and the inside of the cell. A mechanistic explanation of this phenomenon can be found as follows:

Action potential: a mechanistic explanation of this phenomenon specifies parts such as voltage-gated sodium and potassium channels. It describes how activities of the parts like influx and efflux of ions through the channels underlie the rapid changes in membrane potential. In this case, mechanistic explanations spell out the relevant physical details. Hodgin and Huxley model is a major achievement that is not a mechanistic explanation of the action potential.

Marraffa and Paternoster (2013: 14) describe well how the account of explanatory integration given by Craver (2007) and colleagues such as Becthel (2009), entails a sort of "inter-level" mechanistic explanations. To spell out the notion of explanatory integration, Craver (2007) examines the development of the explanations of Long-Term Potentiation (LTP) and spatial memory. He distinguishes at least four levels. At the top of the hierarchy (the behavioral-organismic level) are memory and learning, which are investigated by behavioral tests. Below that level is the hippocampus and the computational processes it is supposed to perform to generate spatial maps. At a still lower level are the hippocampal synapses inducing LTP. And finally, at the lowest level, are the activities of the molecules of the hippocampal synapses underlying LTP (e.g., the N-methyl Daspartate receptor activating and inactivating). These are "mechanistic levels" or "levels of mechanisms": the N-methyl D-aspartate receptor is a component of the LTP mechanism, LTP is a component of the mechanism generating spatial maps, and the formation of spatial maps is a part of the spatial navigation mechanism. Integrating these four mechanistic levels requires both a "looking up" integration, which will show that an item (LTP) is a part of a upper-level mechanism (a computational-hippocampal mechanism); and a "looking down" integration, which will describe the lower-level mechanisms underlying the higher-level phenomenon (the molecular mechanisms of LTP)". According to mechanists, this account is well-suited to define the explanatory integration in Cognitive sciences. Due to the redefinition of explanations under the label of information concerning mechanisms involved in cognitive phenomena, this kind of explanation gives rise to what Khalifa et al. (2022) calls the Mechanistic-Based Integration of explanations in Cognitive sciences.

3.2   Computational

Most prominent alternative to mechanistic explanations in the philosophical literature: they are considered a subset of functional explanations–explain phenomena by appealing to their function and the functional organization of their parts (Fodor 1968; Cummins 1975, 1983, 2000) The functions to which they appeal involve information processing. In computational explanations, a phenomenon is explained in terms of a system performing a computation. A computation involves the processing of input information according to a series of specified operations that results in output information. Many computational explanations describe the object of computation as having representational content, but some challenge this as a universal constraint on computational explanations.

3.3   Topological

In topological explanations, a phenomenon is explained by appeal to graph-theoretic properties. Scientists infer a network's structure from data and then apply various graph-theoretic algorithms to measure its topological properties, which are structural or mathematical properties of the system. In contrast to mechanistic explanations, they abstract away from particular details of causal interactions or mechanisms found in the phenomenon: "For instance, clustering coefficients measure degrees of interconnectedness among nodes in the same neighborhood. Here, a node's neighborhood is defined as the set of nodes to which it is directly connected. An individual node's local clustering coefficient is the proportion of edges within its neighborhood divided by the number of edges that could possibly exist between the members of its neighborhood. By contrast, a network's global clustering coefficient is the ratio of closed triplets to the total number of triplets in a graph. A triplet of nodes is any three nodes that are connected by at least two edges. An open triplet is connected by exactly two edges; a closed triplet, by three. Another topological property, average (or "characteristic") path length, measures the mean number of edges needed to connect any two nodes in the network" (Khalifa et al. 2022: 6). This kind of explanation is used to picture how some system has the property to efficiently propagate information, as the case study of the nervous system of Caenorhabditi elegans show (Watts and Strogatz 1998; Latora and Marchiori 2001; Bullmore and Sporns 2012).

3.4   Dynamical

When we have dynamical explanations, phenomena are accounted for using resources of dynamic system theory. A system is dynamical if its state space can be described using differential equations. The equations describe the evolution of the system over time. For example, in the case of dynamic explanations of bimanual coordination, the explanation rests on the fact that only the in - and - anti - phase oscillations of the index fingers are basins of attraction. Also this kind of explanation, while picturing some dependency relations between the features of the phenomenon, is not of the mechanistic kind.

4   Scientific Understanding and Explanatory Integration

Scientific understanding is a multifaceted and dynamic concept that plays a central role in the field of cognitive science. The intricate workings of the human mind demand an array of explanations that span across disciplines, encompassing neuroscience, psychology, philosophy, and artificial intelligence. What emerges from the case studies in cognitive science is a compelling argument: scientific understanding can serve as a unifying framework that harmonizes the diverse kinds of explanations inherent in this multidisciplinary field. This notion of scientific understanding does not seek to reduce explanations to a singular, reductive framework but rather embraces the plurality of explanations, offering a holistic and integrative perspective. In examining the features of this integrative scientific understanding, we find that it is characterized by its depth, coherence, pragmatism, and its ability to promote interdisciplinary collaboration, ultimately enriching our comprehension of cognitive phenomena.

Khalifa et al. (2022) propose two main ways to integrate different kind of explanations in cognitive sciences: the Understanding-Based Integration (UBI) and the Mechanism-Based Integration (MBI). UBI is ultimately a new view about explanatory integration in Cognitive sciences, while MBI concerns the received view about explanatory integration aiming at unifying the different levels of explanation in a mechanistic one. According to Taylor (2021) we should not accept to dismiss cross-explanatory integrations of mechanistic, dynamicist, psychological, computational and topological explanations in cognitive sciences, as instead some philosophers argue (Kaplan and Carver 2011; Miłkowski 2016; Piccinini and Craver 2011). Khalifa is also a defender of pluralism in cognitive sciences regarding explanatory integration and he argues for an account of explanatory integration based on SU. On the other hand, MBI provides that all models in the cognitive sciences are explanatory only insofar as they give information about mechanistic explanations. Against this, defenders of pluralism provide examples of putatively non-mechanistic explanations. In response, MBI philosophers use two strategies. The negative strategy consists in revealing that the putatively non-mechanistic explanation are no explanation at all (Kaplan 2011; Kaplan and Craver 2011). The other strategy is assimilation and reveals the putatively non-mechanistic explanation to be a mechanistic explanation but with an elliptical nature (Piccinini 2006: 205, Piccinini and Craver 2011, Miłkowski 2013; Povich 2015; Hochstein 2016). In recent years another area of inquiry has gained a central protagonism also in cognitive science, namely the study of the family of algorithms collected under the umbrella name "deep-learning". These algorithms are run on machines to reach human-like abilities in many tasks. As we have seen in Chap. 5, deep learning models are emerging from the connectionist paradigm and are now basically studied for engineering purposes, but they seem to be useful also for cognitive aims. According to Perconti and Plebe (2020), deep learning models pose questions that cognitive science should try to answer, such as why deep convolutional models that are disembodied, inactive, static and free of contextual awareness, seem to be the closest representation to the patterns of activation in the brain visual system (Perconti and Plebe 2020). They argue that deep learning "can and should have its say in cognitive science" on the basis that "the engineering objectives of deep learning have been met with such success that, for the first time, we have artificial models performing complex cognitive tasks at human performance level. The era of toy worlds in which models are restricted to highly simplified versions of cognitive capabilities is over. We now have empirical examples of algorithms solving cognitive tasks at the full scale of complexity" (Perconti and Plebe 2020: 2). The question is: can Khalifa's framework account for this turn?

The main flow I detect in Khalifa's framework to this extent is that the EKS model can account for the explanatory knowledge in the process of achieving understanding - and it also succeeds as an account of explanatory integration, when the items in question are explanations consistent with a broad theory of explanation Khalifa is entitled to. In the example suggested by Perconti and Plebe, we can figure out a case in which deep learning models of a phenomenon play the role of a representation of such phenomenon, and due to their features, researchers can achieve a representational understanding of the target-system, i.e. the patterns of activation in the brain visual system. We could extract explanatory information from the deep learning models in question, but according to Khalifa's framework, the information will be not enough to account for explanations. Still, the models can play a relevant role in the process of understanding phenomena, even if the explanatory information researchers will gain from them is at minimum in comparison to the use of different experimental methodologies.

On one hand, we have Khalifa's account, which is broad enough to account also for the case of explanatory integration in the cognitive science. On the other hand, I still feel the pressure–if we want an account of scientific understanding sound enough to capture also what happens in the scientific research using deep learning models–not to limit the epistemic account of SU to explanatory knowledge. Relevant information is obtained from researchers through representations of the target system, developed through deep learning models. So, if we want to capture also this cases of SU in cognitive science, some other conditions must be put in play. Here I recognise the value of Khalifa and colleagues' argumentation concerning the analysis of explanations and the consistency of their view with explanatory integration in the cases they study, although I suggest that the following conditions have to be satisfied, in order to account for both explanatory and representational understanding: Understanding Pluralism, Coherence Across Explanations, Pragmatic Utility, Interdisciplinary Collaboration, and Non-Reductive Nature of Understanding. I sketch these conditions in the following lines, suggesting that their satisfaction could improve Khalifa and colleagues' account of UBI, even if they leave some open problems.

4.1   Understanding Pluralism

As demonstrated in Cognitive sciences case studies, scientific understanding delves beyond surface-level knowledge. It encompasses the ability to penetrate the layers of causality, mechanisms, structures, and functions that underpin cognitive processes. For instance, when exploring the concept of mirror neurons, scientists go beyond the mere awareness of their existence and investigate the neural mechanisms (causal-mechanical explanations), how they relate to imitative behaviours (structural explanations), and why they evolved (functional explanations). One could argue that in each explanatory setting, a distinct species of understanding is achieved so that we can have, for example, a local understanding concerning the phenomenon related to the neural mechanism. Each area of cognitive sciences would then be entitled to achieve a distinct scientific understanding of the phenomena under scrutiny, given the specific explanatory information the scientific research produces locally. With "local understanding" I want to describe the case in which cognitive scientists gain a complete insight into the phenomena they study, providing a richer and more nuanced perspective than mere factual knowledge. The locality here is designed by the specific features of the scientific research in the distinct area. In this way, we should end up not only with different kinds of explanations but with different kinds of local understanding specific to each area, and different understandings for different areas of cognitive sciences.

Can this pluralism of understanding be integrated in the same way we want to gain explanatory integration? As the plurality of explanations has been tackled by two mechanistic and understanding frameworks, can we account similarly for the plurality of understanding? Someone could argue that each area of inquiry in the cognitive sciences does not give rise to an instance of scientific understanding. This would obliterate the issue of accommodating different kinds of understanding. But if the EKS model captures the instance of understanding in a specific context, with defined explanations at disposal, I submit that in each area of inquiry, explanations come with a distinctive local understanding. This means that we also need a way to integrate the local understanding, not only with the explanations but with the general and broader scientific understanding, which is the result of the explanatory integration. It will be a pluralistic understanding in two ways: the first is that it must account for the local levels of understanding, and the second is that it concerns the plurality of explanations we want to integrate together. The relation between locals and general understanding should be explored in more detail, as its interplay with integrated explanations. To conclude, it is important to fill this gap in Khalifa and colleagues' account of explanatory integration in cognitive sciences to stabilise the definition of scientific understanding in this interdisciplinary research area.

4.2   Coherence Across Explanations

One distinguishing feature of integrative scientific understanding is its capacity to weave together disparate threads of explanation into a cohesive tapestry. Rather than isolating causal-mechanical, structural, and functional explanations, it seeks to align and integrate them. In doing so, it connects the dots and identifies the points of convergence and divergence within the explanations. This coherence fosters a more comprehensive and interconnected view of cognitive phenomena, highlighting the intricate relationships between different facets of understanding. Given the many kinds of explanations involved, the UBI proposed by Khalifa and colleagues should be able to satisfy the requirement of being coherent. It is a well-debated issue (Khalifa 2016) in the epistemology of coherence and understanding. It is related to the coherence justification epistemology. On this theme, Elgin (2007, 34) writes: "[An individual] proposition derives its epistemological status from a suitably unified, integrated, coherent body of information. This is the core conception of understanding [...] And it is the conception of understanding that is closely connected to explanation". And also Kvanvig (2003, 192) says: "The central feature of understanding, it seems to me, is in the neighborhood of what internalist coherence theories say about justification. Understanding requires the grasping of explanatory and other coherence-making relationships in a large and comprehensive body of information". They are the principal defenders of the strong view about coherence in understanding. Khalifa (2016, pagina) on the other side argues that the relation between understanding and coherence is shallow: " coherence is not part of the "core conception of understanding." Similarly, while the "central feature of understanding" is in the neighborhood of coherence, it isn't at home there. On my view, understanding is quasi-coherent: it walks like coherence and talks like coherence, but does not require a coherentist epistemology". According to Khalifa (2016), the improvements in understanding are not due to coherence, which is implied in the objectual understanding and not in the explanatory understanding. Still, the issue of a definition of coherence among the explanations, given their different journeys through the sciences remains one to be tackled.

4.3   Pragmatic Utility

Scientific understanding, in its integrative form, is pragmatically useful. The idea that understanding is a pragmatic notion is already embedded in De Regt's account, while Khalifa sides with the received view of understanding, conceiving it as bound to explanatory information. According to them, scientific understanding is not a purely theoretical construct but rather a skill or ability that aids researchers in making predictions, explaining observations, and solving complex problems. Considering the study of working memory and executive function, we can see that integrating insights from neuroscience, cognitive psychology, and artificial intelligence enables researchers to develop practical models that simulate and predict these cognitive processes. The scientific understanding scope is then not only determined by the phenomena under research, but its applications can range different scientific areas. Scientific understanding in this way is not only an ability, but it has also a pragmatic utility, which broaden the specific scope of understanding a specific phenomenon or problem. This pragmatic utility not only deepens our understanding but also allows for the application of cognitive science findings in practical domains like education, healthcare, and technology development. I think that Khalifa and colleagues' account of UBI should tell more about the pragmatic utility of local and general instances of scientific understanding.

4.4   Interdisciplinary Collaboration

Perhaps one of the most striking features of this form of scientific understanding is its ability to foster interdisciplinary collaboration. In the case studies mentioned earlier, the integration of neuroscientific, psychological, and computational explanations exemplifies how cognitive scientists from diverse backgrounds can come together to tackle complex problems. The exchange of insights and methodologies across disciplines enriches the overall understanding of cognitive phenomena. Moreover, it encourages researchers to embrace the diversity of explanations, recognizing that different disciplines bring unique perspectives and tools to the table. The program of establishing the UBI framework for cognitive sciences should also delve into the sociology of science, since the collaboration among researchers and scientific communities is at the core of the possibility of explanatory integration.

4.5   Non-reductive Nature

Importantly, this kind of scientific understanding is not reductive. It does not seek to reduce complex cognitive phenomena to a singular, oversimplified explanation. Instead, it acknowledges the multiplicity of factors and dimensions that contribute to our comprehension of these phenomena. While it integrates diverse explanations, it does so in a way that respects the complexity and richness of cognitive science, recognizing that no single explanatory approach can capture the entirety of the field. While scientific understanding comes with a context-sensitive nature, recognized also by Khalifa, knowledge as a de-contextualising device to structure information in a coherent, justified and approximately true form. Understanding is then a tool to get knowledge and to foster scientific knowledge in many areas of inquiry.

To conclude, scientific understanding in cognitive science provides a framework that unifies and integrates the diverse kinds of explanations inherent in this multidisciplinary field. It is marked by its depth, coherence, pragmatic utility, and capacity to promote interdisciplinary collaboration. This form of understanding does not seek to reduce cognitive science to a singular explanation but rather embraces the plurality of explanations, enriching our comprehension of the intricate workings of the human mind. It is a testament to the dynamic and evolving nature of scientific understanding, which continues to drive progress and innovation in the field of cognitive science. Given the fruitful connections made by Khalifa and colleagues, they should specify whether according to their view, UBI does or does not lead to new scientific knowledge. If the aim of explanatory integration is to ensure and expand the relevant scientific knowledge of cognitive phenomena, UBI should play an important role in affirming it.

5   Conclusion

The explanatory integration in cognitive sciences is a relevant issue, not only in the specific domain of cognitive researchers, but also for philosophers working on the notion of understanding and in particular scientific understanding.

The UBI framework is promising and it depicts a possible development for the understanding studies, conceived more broadly as a philosophical and cognitive endeavour.

In conclusion, scientific understanding in cognitive sciences requires the integration of diverse explanations, as defended by Khalifa et al. (2022). Khalifa's account of explanatory scientific understanding provides a valuable framework for achieving this integration. Still, their program lacks some clarifications about the relation between local and general understanding, the coherence of explanations coming from different scientific disciplines and the social interplay between research communities.

Moreover, it has to be emphasized the distinction between understanding and knowledge. The notion of integrative explanations in the scientific understanding framework highlights the need for a broader description of cognitive phenomena. By embracing an interdisciplinary approach and showcasing case studies, this paper advocates for a more unified and structured perspective on scientific understanding in cognitive sciences, ultimately advancing our understanding of the human, animal and artificial mind.

References

Bechtel, W., Richardson, R.C.: Discovering complexity: Decomposition and Localization as Strategies in Scientific Research. Princeton University Press, Princeton (1993)

Bullmore, E., Sporns, O.: The economy of brain network organization. Nat. Rev. Neurosci. **13**, 336–349 (2012)

Craver, C.F.: Beyond reduction: Mechanisms, multifield integration and the unity of neuroscience. Stud. Hist. Philos. Sci. Part C Stud. Hist. Philos. Biol. Biomed. Sci. **36**(2), 373–395 (2005)

Craver, C.F.: Explaining the Brain: Mechanisms and the Mosaic Unity of Neuroscience. Clarendon Press, Oxford (2007)

Cummins, R.C.: Functional analysis. J. Philos. **72**, 741–765 (1975)

Cummins, R.C.: The Nature of Psychological Explanation. MIT Press, Cambridge, MA (1983)

Cummins, R.C.: "How does it work?" versus "What are the laws?": Two conceptions of psychological explanation. In: Keil, F.C., Wilson, R.A. (eds.) Explanation and Cognition, pp. 117–144. The MIT Press, Cambridge, MA (2000)

De Regt, H., Dieks, D.: A contextual approach to scientific understanding. Synthese **144**, 137–170 (2005)

De Regt, H.: Understanding Scientific Understanding. Oxford University Press, Oxford (2017)

Elgin, C.: Understanding and the facts. Philos. Stud. **132**, 33–42 (2007)

Fodor, J.: A. Psychological Explanation: An Introduction to the Philosophy Of Psychology. Random House, New York, (1968)

Gentner, D.: Cognitive science is and should be pluralistic. Top. Cogn. Sci. **11**(4), 884–891 (2019)

Kaplan, D.M.: Explanation and description in computational neuroscience. Synthese **183**, 339–373 (2011)

Kaplan, D.M.: Explanation and Integration in Mind and Brain Science, 1st edn. Oxford University Press, Oxford (2017)

Khalifa, K.: Inaugurating understanding or repackaging explanation? Philos. Sci. **79**, 15–37 (2012)

Khalifa, K.: Is understanding explanatory or objectual? Synthese **190**, 1153–1171 (2013a)

Khalifa, K.: The role of explanation in understanding. Br. J. Philos. Sci. **64**, 161–187 (2013b)

Khalifa, K.: Must Understanding be Coherent? In: Grimm, S., Baumberger, C., Ammon, S. (eds.) Explaining Understanding, pp. 139–165 (2016)

Khalifa, K.: Understanding, Explanation, and Scientific Knowledge. Cambridge University Press, Cambridge (2017)

Khalifa, K.: Is Verstehen scientific understanding? Philos. Soc. Sci. **49**, 282–306 (2019)

Khalifa, K., Islam, F., Gamboa, J.P., Wilkenfeld, D.A., Kostic, D.: Integrating philosophy of understanding with the cognitive sciences. Front. Syst. Neurosci. **16**, 1–17 (2022)

Khalifa, K.: Should friends and frenemies of understanding be friends? discussing de Regt. In: Khalifa, K., Lawler, I., Shech, E. (eds.) Scientific Understanding and Representation: Modeling in the Physical Sciences. Routledge, London (2023a)

Khalifa, K., Lawler, I., Shech, E.: Scientific Understanding and Representation: Modeling in the Physical Sciences. Routledge, London (2023b)

Knavig, J.: The Value of Knowledge and the Pursuit of Understanding. Cambridge University Press, Cambridge (2003)

Latora, V., Marchiori, M.: Efficient behavior of small-world networks. Phys. Rev. Lett. **87**, 198701 (2001)

Legrenzi, P.: Prima lezione di scienze cognitive. Roma, Laterza (2002)

Marconi, D.: Filosofia e Scienza cognitiva. Roma, Laterza (2001)

Miłkowski, M.: Explaining the Computational Mind. Massachusetts, MIT Press, Cambridge (2013)

Miłkowski, M.: Unification strategies in cognitive science. Stud. Logic Gramm. Rhetor. **48**(1), 13–33 (2016)

Neurath, O.: Unified science and its encyclopaedia. Philos. Sci. **4**(2), 265–277 (1937)

Oppenheim, P., Putnam, H.: Unity of science as a working hypothesis. Minn. Stud. Philos. Sci. **2**, 3–36 (1958)

Perconti, P., Plebe, A.: Deep learning and cognitive science. Cognition **203**, 104365 (2020)

Piccinini, G., Craver, C.: Integrating psychology and neuroscience: functional analyses as mechanism sketches. Synthese **183**, 283–311 (2011)

Watts, D.J., Strogatz, S.H.: Collective dynamics of 'small-world' networks. Nature **393**, 440–442 (1998)

Code Biology and Enactivism: Bringing Adaptors to Basic Minds

Rasmus Gahrn-Andersen[(✉)] [iD]

University of Southern Denmark, 4200 Slagelse, Denmark
rga@sdu.dk

Abstract. The paper presents a preliminary attempt at exploring the potentially fruitful intersection between Code Biology and enactivist research. As an umbrella label, 'Enactivism' designates different strands of research in the radical, anti-representationalist cognitive sciences. The paper begins by motivating the need for nuancing basic claims central to Autopoietic Enactivism and makes the argument that Code Biological insights and notions can be used to deflate the supposedly strong autonomy of individual sense-makers. Then, it goes on to show that Code Biology has the potential for informing Sensorimotor Enactivism by revealing some of the complex codifying mechanisms involved in not just visual perception but virtually any kind of sense-making across the evolutionary board. The paper builds on these insights to substantiate Radical Enactivism's information-as-covariance and, beyond this, the mechanisms which enable 'basic mentality'. Finally, a critical developmental perspective is considered.

Keywords: Cognition · Basic minds · Adaptors · Ecological codes · Life-mind continuity

1 Introduction

According to Barbieri, Code Biology can be understood in two senses. In its specific sense, Code Biology is the study of organic codes across the scales of evolution and in relation to the biological functioning of living beings whether these belong to Archaea, Bacteria or Eukarya [1]. As such, Code Biology in the specific sense can be traced back to the middle of the last century with the discovery of the genetic code [2] and the subsequent discovery of the epigenetic code [3], the histone code [4] etc. In its more general sense, Code Biology is the "study of all codes of life, from the genetic code to the codes of culture" [1] and it was introduced by Barbieri [5]. In its general sense, Code Biology comprises "a deeply interdisciplinary field and necessarily requires the contribution of different scholars: biologists, neuroscientists, ecologists, linguists, philosophers, mathematicians and computer scientists" [1] (p. 4). Whether it is understood in the specific or general sense, Code Biology explores codified relations not just by means of experimental work in accordance with standard scientific methods but also theoretically [6] (p. 2). It comprises an alternative to the chemical paradigm in biology and, hence, the assumption that the genetic code is characterized by fixed translations from sequences of codons into proteins meaning that a codon is bound to translate into particular amino acids [7] (p. 14). Conversely, a Code Biology framework takes stock of the fact that the genetic code is governed by arbitrary rules that are not inscribed with physical or chemical necessity thus recognizing that the same sequence of codons might translate into different proteins [6] (p. 2). In terms of definition, a code is a "mapping between the objects of two independent worlds that is implemented by the objects of a third world called adaptors" (Barbieri [8]–quoted in [6] (p. 6)). This very broad construal of the ontology of a code entails that Code Biology can be viewed as a general framework for not just encompassing codes pertaining to the biological realm but also as extending into the neural and the cultural domain (cf. [9]).

The current paper presents a first, preliminary attempt at discovering the possible synergies between Code Biology in the general sense (Sect. 1.1) and Enactivism. Enactivism is another general position which seeks to investigate the link between life and mind (broadly understood) and, hence, everything between the single cell and human cultural practices. Although having a strong basis in theoretical biology, Enactivism is predominantly a strand of research that has evolved in the context of the so-called radical cognitive sciences. Anti-representationalist models describe cognition as simpler processes compared to proponents of representationalism, who consider mental content as playing an enabling, intermediate role between, on the one hand, an agent's cognitive and behavioral states, and, on the other, the environment. Also, anti-representationalist research investigates the embodied and embedded aspects of cognition (including affordance use), thus taking extrinsic factors into account, which are typically not considered by traditional cognitivist approaches, given their neurocentric focus. The 'Enactive approach' was born with the publication of The Embodied Mind in 1991. Here, Varela and colleagues make the programmatic claim that "cognition is not the representation of a pregiven world by a pregiven mind but is rather the enactment of a world and a mind on the basis of a history of the variety of actions that a being in the world performs" [10] (p. 9). The view challenges orthodox cognitive science and so-called 'cognitivist models' which, amongst other things, treat the mind as a digital computer involving planning based on symbol processing [11]. Thus, as Barandiaran [12] argues, enactivists challenge the idea that cognition unfolds as the intermediate step between sensed input and behavioral output as they assume a dynamic perspective on how the mind unfolds (cf. p. 410). As Hutto and Myin phrase it, enactivists thus hold "that the embedded and embodied activity of living beings provides the right model for understanding minds" and, hence, that "[t]o understand mentality, however complex and sophisticated it may be, it is necessary to appreciate how living beings dynamically interact with their environments" [13] (p. 4). Yet, as a position, enactivism is becoming increasingly heterogeneous with about a handful of different approaches that all commit to basic enactivist principles concerning the need for overcoming the computer-metaphor when explaining cognition. Consequently, I limit the inclusion to three of the most prominent branches namely Autopoietic Enactivism (Sect. 2), Sensorimotor Enactivism (Sect. 3) and, finally, Radical Enactivism (Sect. 4). The purpose in so doing is not only to stress compatibility but also to explore what a Code Biology-informed take on enaction potentially has to offer. The paper concludes with some critical remarks concerning the current state of Code Biology and its possible prospects in an enactivist context (Sect. 5).

1.1 Code Biology (In the General Sense)

In the general sense, Code Biology stipulates a 'three worlds' view. As Barbieri writes, Code Biology, in the specific sense, is dedicated to exploring phenomena unfolding strictly in the organic world (World 1), where no appeal to 'interpretation' is needed to make sense of them. In other words, the processes unfolding in this particular realm can be exclusively understood in terms of 'organic semiosis' and the organic mechanisms that enable them [1]. However, animal and human semiosis (Worlds 2 and 3) are assumed to be different in the sense that they not only involve organic processes but also neural ones. Thus, Barbieri claims that an appeal to organic mechanisms cannot stand alone. It must also include, at the very least, so-called 'interpretive' ones. In the case of humans, there is an additional component that is characteristic of its functioning, namely a reliance on symbols. Whereas other types of signs (i.e., icons and indexes) presuppose a sort of natural connection between a sign and its object in terms of either structural similarity or a physical link, symbols differ [7] (p. 186). Symbols are manmade and relative to the specific conventions and rules pertaining to certain socio-cultural practices. As such, symbols are signs characterized by an arbitrary relation between the sign itself and its object or referent. For instance, a word is a symbol because there is no natural connection between a word ('cat') and its object/referent ('a furry four-legged animal that purrs'). The relation between the two is necessitated merely by social conventions.

Yet, when it comes to interpretative mechanisms, it is clear that Barbieri assumes that cognition (understood as 'interpretation') is bound to be representational. For instance, at one point, he claims that semiosis in animals occurs as they 'receive signals from the environment, transform them into mental images, and perform mental operations' [7] (p. 143). Elsewhere, Barbieri puts it equally bluntly by arguing that "the mind can only act on representations of the world, and that is why it must use signs which have both internal and external meanings…" (p. 165). However, Barbieri operates with a very broad definition of code from the outset, which makes no ontological claims about the so-called 'worlds' it supposedly connects. Thus, in the context of this paper, it becomes prudent to ask whether Code Biology, in its general sense, can be developed through an interplay with Enactivism to acknowledge anti-representationalist aspects of cognition.

2 Autopoietic Enactivism

The 'Enactive approach' as formulated by Varela and colleagues has effectively evolved into what is now commonly referred to as Autopoietic Enactivism (cf. [14]–for some examples, see [15–18]). Inspired by Maturana and Varela's [19] notion of autopoiesis, Autopoietic Enactivists take the autonomous agency of the individual cognizer as the basis for the evolution of multi-agent systems and all the way up to human culture and language [20] (p. 4). Indeed, as in the case of an autopoietic system, the autonomous agent is 'operationally closed'. This means that the agent is essentially self-constituting with regards to its own identity whereby it enables the agent to differentiate itself over against the environment [20]. The autonomy of agents–which in the minimal sense pertains to the individual cell–comprises the basis for other core enactivist concepts including adaptivity, agency, mentality, sociality and culture [20] (p. 4). In this connection, mentality refers to cognition that is irreducible to metabolic states but also relies on so-called "other-related concerns" such as social norms. Autopoietic Enactivism stresses so-called strong life-mind continuity meaning that "Mind is life-like, and life is mind-like" [15] (p. 385). Living agents are per definition cognizers–or, in autopoietic enactivist parlance: sense-making agents:

> Sense-making is tantamount to cognition, in the minimal sense of viable sensori-motor conduct. Such conduct is oriented toward and subject to signification and valence. Signification and valence do not pre-exist "out there," but are enacted or constituted by the living being. Living entails sense-making, which equals cognition [15] (p. 387).

Effectively, it is agents and their distinctive selves which give rise to their worlds. Sense-making is a radical notion which places emphasis on the constitutive relations brought forward by agents as they engage with their surroundings. Thus, on this view, cognition is inseparable from perception and action; or enacted perception (ibid.). So, although Froese and Di Paolo stress that emergent properties that arise on the basis of the sense-making of autonomous agents also have the potential to affect their own constitutive background [20] (p. 4), the core of cognitive phenomena (also in the sense of whatever precedes basic sense-making) are de facto traceable to autonomous agents and, hence, the minimal case of the autopoietic cell. On this view, meaning or 'significance' is not represented, rather it is dynamically enacted as an organism responds to its environment [11]. In this connection, it is worth noting how Code Biology seems to be able to inform Autopoietic Enactivism on two related fronts:

First, there is the basic insight that life precedes the autopoietic cell. Following Barbieri, Code Biology recognizes that the genetic codes in cells are traceable to a common ancestor (in the form of a ribonucleoprotein system). In other words, such codes exist prior to the evolution of cells where spontaneous proteins and genes prevailed [6] (p. 6). Yet, these lacked specificity in that ancestral systems did not reproduce in the autopoietic sense (and were not individuals) but, instead, gave rise to qualitatively different systems [21] (p. 297). As Barbieri puts it: "Autopoiesis, in other words, did not exist before the first cells, so it was not the mechanism that gave origin to them." [6] (p. 7). Thus, rather than assuming that autopoiesis is intrinsic to life, it makes sense to embrace the view that what Barbieri calls *codepoiesis* precedes autopoiesis and, hence, that there is more to life than the operational closure of living entities. Indeed, the common ancestral system had the sole capacity of generating new codes. This situation changed with the origin of the genetic code which, on the one hand, fixed the rules so that "no other modification in the coding rules was accepted" while, on the other hand, kept a capacity for exploring new codes (ibid.). Bacteria lived for billions of years without becoming multicellular, functioning as individuals, or using vertical reproduction. Since they did rely on DNA, autopoiesis can be seen as having evolved from codepoiesis as a simplex way of limiting adaptivity. As Barbieri [21] summarizes, codepoiesis thus entails the dual processes of code generation and code conservation (p. 298). The view also allows for the fact that not all cells following the evolution of the genetic code are autopoietic in the strict sense of the term (cf. Ibid.). Luisi [22] points out that the same can be said about viruses for the simple reason that a virus neither produces its own protein coat nor its nucleic acids (p. 51). In other words, it is not autopoietic or self-reproducing.

Second, the Autopoietic Enactivist emphasis on the strong autonomy of individual sense-makers is far from unproblematic. In what follows, I summarize an enduring criticism [23–26]. The issue with strong autonomy is exemplified by De Jaegher and Di Paolo [27] who explain the progression from individual sense-making to participatory sense-making by means of a tacit shift in their explanatory locus: whereas individual sense-making is explained as being determined by intrinsic teleology and, hence, agent-internal values, participatory sense-making entails that the strong autonomy of agents is superseded by 'the interaction' in which the agents participate. Indeed, De Jaegher and Di Paolo recognize this when arguing that "not only must the process itself enjoy a temporary form of autonomy, but the autonomy of the individuals as interactors must also not be broken (even though the interaction may enhance or diminish the scope of individual autonomy)" [27] (p. 492).

Critically, this brings about an explanatory issue. For by presupposing that the interaction can take up autonomy in its own right, De Jaegher and Di Paolo tacitly acknowledge that there are circumstances where interactional outcomes do set aside (or: transcend) the strong autonomy of individual agents as the agents take on agent-external norms for guiding their behavior. But it remains unclear to what extent this is at all possible without at the same time violating the strong autonomy of agents that participate in the interaction and, hence, the principle of intrinsic teleology. As an explanatory principle, strong autonomy leaves no room for weak autonomy thus rendering it unclear how agents can operationalize agent-external values and norms (so-called 'culturally prevalent behavioral norms' [27] p. 495) without violating Autopoietic Enactivism's core claim that agents, per default, are strongly autonomous. Autonomy can be traced to the very core of the notion of 'enact' in the sense of 'acting from within' [21]. Another way of conceiving this problem is to see it as being related to how autonomy and heteronomy are both influential in an agent's sense-making activities. Relatedly, some scholars such as Kirchhoff [28, 29] have criticized autopoietic theory for the fact that it offers internalist explanations, arguing that adaptivity–a crucial aspect of life-mind continuity–cannot be adequately explained unless one instead adopts an externalist explanation like the one linked to the so-called 'Free Energy Principle' [28] (p. 2363). Yet, as Allen and Friston [30] show, proponents of the Free Energy Principle assume that cognition relies on a degree of internalism and more specifically, prediction which ties with autopoiesis which, on their view, is not entirely internalist (as presented Kirchhoff) but nevertheless entails a degree of 'extended' autopoiesis which involves aspects of the world as well as those internal to the agent. For as they claim, the "causal machinery of the brain and its representations are enslaved within the brain-body-environment loop of autopoiesis" (p. 2477).

Although recognizing that fully developed cells indisputably come with autopoietic traits, Code Biology is not tied to a commitment to strong autonomy. Thus, it has shown to have no problems in recognizing the fact that cognition can be both shaped by extrinsic and intrinsic factors. For instance, as to the ontology of code, it is worth noting that it is indeed not anchored in individual agents or organisms. Rather, it is ontologically unspecific in the sense that it is open to interpretation of what counts as 'independent worlds' and, further, what characterizes their differences as such. Such lack of specificity might not be a bad thing since it provides ample room for the development of Code Biology in the general sense. For instance, Gahrn-Andersen and Prinz have stressed how prosthetic device-usage showcases the fact that human agents are weakly autonomous and, hence, the fact that the heteronomy of social norms and values play a constitutive role to the successful integration of a prosthesis [26]. Specifically, they argue that the experience of being 'whole' which is a prerequisite for optimal performance in any skillful activity is secured only if the prosthetics wearer accommodates their sensorimotor contingencies to the embodied actions recognized by a community. In this connection, heteronomy is indispensable because "the agent would have to adapt to existing norms and rules regarding how particular movements should be executed. This is a prerequisite in order for the prosthesis to become fully integrated into the patient's life. Thus, pointing, stirring, clapping, cheering, punching etc. are all actions that are defined by social practices or 'culture' in the sense that they are arbitrarily determined by certain social conventions" [26] (p. 16). On this view, qualitatively different codified relations enable the embodied agency of amputees to re-enter into social communities by, above all, accustoming themselves with their prothesis to the extent that they are able to experientially transcend its presence. Such transcending allows the prothesis to function in a smooth way in the sense that it becomes functionally equivalent to the body part it has replaced.

3 Sensorimotor Enactivism

As introduced by O'Regan and Noë, Sensorimotor Enactivism challenges neurocentrism and the idea that cortical maps can be the sole source for explaining visual perception [31] (p. 393) in the sense that visual experience arises as visual stimuli basically trigger brain-internal mechanism. Rather, proponents of this enactivist branch appeal to the laws of so-called sensorimotor contingencies or, as formulated by Ward and colleagues, 'patterns of dependence obtaining between perception and exploratory activity' [11] (p. 371). These contingencies comprise 'various motor actions' and can thus be used for distinguishing vision from other perceptual modalities [31] (p. 941). Indeed, the key differences between different senses can be traced to the structures of their related actions in the sense that, for example, eye-movements and blinks have an effect on visual perception but not on olfactory or auditory perception (cf. Ibid.). Thus, perception is ultimately based in active, environment-exploration through law-governed contingencies which are relative to one or more of the five senses. In terms of codified activity, Sensorimotor Enactivism hereby brings an important insight to the fore by showing that visual experience is irreducible to neural codes and, thus, how the brain may or may not represent objects [31] (p. 942). Rather, they show that much pertains to the constitution of perceived objects and, more specifically, the contingencies by means of which they are perceived. Thus, the visual system possesses an internal representation that persists across eye saccades thanks to these sensorimotor contingencies. This facilitates the encoding of visual features, thereby enabling perception. The claim gives rise to the somewhat provocative thesis, at least in the context of neural-centric culture, that the neural codes used for representing objects are in fact dependent on the motor actions relative to a given sense (cf. Ibid.). Specifically, O'Regan and Noë argue, they push "the idea that the laws of sensorimotor contingency might actually constitute the way the brain codes visual attributes" (p. 942–943). Thus, on the view of Sensorimotor Enactivism, a sensory modality is "a mode of exploration mediated by distinctive sensorimotor contingencies" [31] (p. 943). Such an ensuring and differentiation of the basic perceptual qualities is insufficient for bringing about perceptual experience. It is also necessary to address the 'awareness' intrinsic to such experiences–an awareness which O'Regan and Noë trace to 'the knowledge' inherent in the sensorimotor contingencies [31] (p. 944). 'Knowledge' thus allows for the skillful execution of a perceptual modality and remains intrinsic to our experience of perceiving a something as a something. O'Regan and Noë explain:

> The experience of red, for example, arises when we know (though this is not propositional, but rather, practical knowledge) that, for example, if we move our eyes over a red region, there will occur changes typical of what happens when our non-homogeneously sampling retinas move over things whose color is red. [31] (p. 963).

Yet, as Noë reveals elsewhere, an appeal to such knowledge is also used by proponents of Sensorimotor Enactivism to stress the continuity between basic perception and more sophisticated or 'intellectual' modes of cognition (e.g., propositional attitudes, planning etc.) [32] (p. 1). Such intellectual attitudes are traditionally deemed to be 'representation-hungry' [33] in the sense of being reliant on mental stand-ins (or: representations). As such, they are normally taken to be the hallmark of human intelligent cognition. In this connection, Noë recognizes that concrete concepts condition human basic perception. Yet, the downside of this move is that such conceptual knowledge conflates with sensorimotor contingencies thus making it difficult to distinguish the two but also to distinguish non-conceptual knowledge from its conceptual counterpart thus leading to an explanatory conundrum. This conundrum emerges from Noë's paraphrasing of Heidegger's point that "the things we encounter are always already familiar" in the sense that it is the basis for such familiarity to arise in the first place. Yet, we find no explanation of how such familiarity arises in the first place given that sensorimotor enactivists place decisive emphasis on prevalent sensorimotor contingencies and, hence, extant knowledge as the enabler of perception [32] (p. 3).

Although Sensorimotor Enactivism has been successful in exploring especially visual perception, it faces a well-known issue namely that it has a narrow focus on human perception in relation to the perceiving human agent and, hence, "largely ignores or downplays the other theoretical principles associated with Enactivism, such as the co-production of organism and environment, emphases on biodynamics and a commitment to life/mind continuity" [11] (p. 371). This resonates with Degenaar and O'Regan's point that Sensorimotor Enactivism restricts its focus to perceptual consciousness [34]. It is especially in this regard that a Code Biology framework has the capacity for informing Sensorimotor Enactivism by revealing some of the complex codifying mechanisms involved in not just visual perception but virtually any kind of sense-making across the evolutionary board.

For instance, Cowley [35] argues in favor of extending the basic principles of the Organic Code-model proposed by Barbieri [7] to include the human domain and, more specifically, the activity of human individuals. In a nutshell, the Organic Code-model describes the basic process wherein protein synthesis unfolds and, hence, how transfer-RNAs (with anchoring RNA) function as adaptors as they relate DNA to ribosomal

RNAs through codified relations. Cowley's basic claim for an extension of the model is that the human body–and herewith what sensorimotor enactivists deem 'sensorimotor contingencies'–also functions in adaptor-like ways [35] (p. 1). Cowley formulates it thus:

> as adaptor-like, humans become code intermediaries who attune to perceived trans-lations at an interface (in Tetris) or in deriving signals from script (in Morse). However, they also develop as apparatuses that can exert whole-body control (based on 'knowing' the rules or 'how to play' the game). Since these powers link skills with expertise, they are cognitive. Even sending signals without under-standing (as in operating in Morse) relies on acting in ways that (inadvertently) prevent and reduce errors. All being well, the system's lee-way enables a person to self-fabricate adaptor-like ways of effective acting. [35] (p. 3).

Cowley contrasts his take on codes with the 'received view of codes' which is prevalent in, for instance, traditional linguistics ('fixed code telementationism' cf. [36]) and representationalism in the cognitive sciences and philosophy of mind and that treats functionality as fundamentally rule-governed and, hence, sense-making processes as passively triggered by means of external stimuli thus eliciting basic linear 'Sense (input) → Plan (compute) → Action (output)'-models. As Barandiaran argues, such models are the negative target of enactivist research [12] (p. 410). Thus, the received view places emphasis on how adaptors use codes irreducible to the rules of the codes themselves thus granting flexibility and unpredictability in code-based processes. On Barbieri's view, however, adaptors bring forth the 'coding rules' [7] (p. 43). They comprise the rules that "establish a mapping between two independent worlds" and, hence, reveal "the presence of a code" [7] (p. 35). Codemakers, on the other hand, ensure the making of the code itself as a kind of artifact-making [7] (p. 12) and, hence, are the agents of coding processes [7] (p. 26):

> Signs, meanings and conventions, however, do not come into existence of their own. There is always an 'agent' that produces them, and that agent can be referred to as a codemaker because it is always the making of a code that gives origin to semiosis [7] (p. 30).

In a crucial addition, Cowley shows how human embodied agents can function both as adaptors and codemakers meaning that adaptors are not subdued or strictly conditioned by codemakers as suggested by Barbieri [35].

Although the relation between the two has yet to be systematically clarified in the realms beyond organic codes, it is evident from Cowley's work that, at least in terms of human meaning-making activities, the wide-view of cognition entails that adaptors and codemakers are basically coinciding. On the one hand, the agent functions as an adaptor in processes defined by external rules (e.g., social rules and conventions). On the other hand, the agent skillfully enacts their understanding of such rules in relation to what they ascribe status to as 'a sign'. The latter effectively brings about not just semiosis but the effectuation of the code. As such, the code is thus realized in the semiotic nexus of, on the one hand, a received or learned code (e.g., the mastering of social conventions or 'adding meaning to information' cf. [8] (p. 95)) and, on the other, the enactment of a relation which builds on the former but is irreducible to it. More specifically, the relation is irreducible to social conventions because it brings about unique experience or meaning thus doing more than simply adding meaning to information. With Cowley's contribution emphasis falls on how agents bring about novelties based on codified processes thus exhibiting creative and adaptive 'self-fabrication' [35] (p. 2).

But Code Biology-based insights can also be used to show the wideness of cognition (to paraphrase Wilson [37]) thus testifying to how it extends beyond the situated individual agent (Cowley's focus) by including populations. Indeed, we must keep in mind that codes also function as 'community rules' [8] (p. 95). For instance, in evoking appeal to soundscapes, understood as 'a vocal milieu', Farina shows how 'semiotic mechanisms' play a decisive role the sense-making of soniferous species [34]. Emphasis is placed on the fact that the capacities for producing and receiving meaningful sounds are ultimately tied to an animal's genome in the sense that "each individual [belonging to different species] has the capacity to extract information from a series of acoustic cues" [38] (p. 150). Yet, Farina also recognizes the vital role played by complex adaptive processes and, specifically, the role of 'cultural transmission' within species communities [38] (p. 148) as well as more general soundscape traits. Farina shows that sensorimotor engagements are irreducible to individuals in that it spreads across populations and the communities they have formed (at the so-called 'acoustic community level') [38] (p. 149). So, although species are said to perceive their environment by means of 'function-specific cognitive templates' [38] (p. 151) based on their needs, it is far from all perceptual processes that are driven by internal mechanisms. While some emphasis is placed on biological codes and autopoietic processes in individual organisms, Farina also stresses the need for considering other ecological codes. In fact, ecological codes including ecoacoustic ones are vital for ensuring organism-environment feedback loops: "[T]he biological codes initiate a process of intra–and interspecific communication, and the ecological codes extend such communication mechanisms to the environment, creating a flow of continuous semiotic feedback between organisms and their habitats" [38] (p. 150).

The presence of such loops is the basic requirement for affecting the genome and, hence, in allowing for population-level patterns to emerge over time or as Farina formulates it: "complex codes operate within specific evolutionary forces that shape the genome and require time to spread and establish across populations" (ibid.). Ecoacoustic events are irreducible to what unfolds on the ecoacoustic community level as they are also relative to the overall soundscape which includes ecological contingencies and human behavior. Crucially, and in parallel to the sensorimotor enactivist focus on the conceptual meanings involved in human perception, the theory of soundscapes showcases how the categorical aspects of conceptual perception are non-conceptually implied in ecoacoustic codes [38] (p. 150) and genetically conditioned in population dynamics.

4   Radical Enactivism

Radical Enactivism or, REC, comprises a radicalized alternative to both Sensorimotor Enactivism and Autopoietic Enactivism. Hutto criticizes how sensorimotor enactivists tend to fall into a cognitivist trap when making claims concerning the brain's role in perception by stressing its capacity to 'judge', 'assume' and 'conclude' [39] (p. 392). Thus, as Hutto states, it very much looks as if perception is taken to involve propositional attitudes and, hence, mental content. So, although O'Regan and Noë are committed to a non-cognitivist take on cognition, it is nevertheless the case that they tend to conflate practical, non-represented knowledge with its propositional, represented counterpart (cf. Ibid.). With regards to Autopoietic Enactivism, REC pushes a similar negative claim although explicitly recognizing that proponents of this particular strand differ from those sympathetic to the sensorimotor branch in that they seek to "make a complete break with cognitivism and representationalism" [13] (p. 33). REC's criticism of Autopoietic Enactivism predominantly relates to the use of metaphors such as 'production' or 'consumption' when proponents of the latter attempt to explain how meaning in sense-making processes is relationally constituted in the coupling of agents and their environment [13] (p. 35). Effectively, this reflects that there is not a decisive move away from cognitivist parlance. Moreover, Hutto and Myin (2013) target the fact that supporters of Autopoietic Enactivism use the same general notions (e.g., 'cognition', 'understanding', 'interpretation') across the board and thus exhibit a "quite liberal understanding of the nature of cognition" (ibid.). In this connection, they argue, it is important to bear in mind that:

The simplest life forms are capable of an intentionally directed responding of a kind that when suitably augmented provides a necessary platform for cognition, interpretation, understanding, sense-making, and emoting; however, their activities do not, in and of themselves, qualify as these forms of mentality. [13] (p. 36).

In effect, on a REC-view, the use of notion such as 'understanding', 'sense-making', 'interpretation' fails to bring clarity to basic modes of cognition because they denote phenomena which are generally not characteristic of basic mentality. As a means for securing REC's radicality, Hutto and Myin express a commitment to the so-called Radical Embodiment Thesis introduced by Chemero [40]. In so doing, they push the view that cognitive behavior which is normally explained by means of computationalist and representationalist models can also simply be explored by means of 'dynamical' explanations (i.e., explanations coming from dynamical systems theory) [13] (p. 2). For as Varela and colleagues [10] also stressed, far from all cognition involves content [13] (p. 5). As a means for countering the shortcoming of Autopoietic Enactivism, REC evokes a distinction between basic forms of cognition and more sophisticated kinds pertaining to humans and which involve 'judging', 'understanding' etc. The difference between the two is ultimately traceable to the fact that they involve two qualitatively different kinds of information: Information-as-covariance and information-as-content [13] (p. 67). With this distinction in place, it becomes possible for REC to acknowledge that more sophisticated, content-involving kinds of cognition differ in kind from their basic counterparts precisely because they involve mental content whereas basic modes involve organic responses which neither "create, carry or consume meanings" [13] (p. 34). Thus, so-called 'basic minds' exploit covariance relations while scaffolded cognition pertaining to certain human-specific phenomena such as 'propositional attitudes', 'language' and 'careful planning' involve mental content [13] (p. xviii, 40, 65).

Basic minds "are fundamentally, constitutively already world-involving" [13] (p. 137) meaning that they are not internal to the cognizer's brain but instead constituted through "concrete patterns of environmental situated organismic activity, nothing more or less" [13] (p. 11). As such, these minds do not exploit 'messages' nor rely on information that is communicated through encoding and decoding sequences [41] (p. 31). Basic minds are exemplified by certain developments in behavioral robotics including, most prominently, the so-called Creatures developed by Rodney Brooks at MIT in the 1980s [13] (p. 41). Brooks pioneered first-generation behavior-based robots by showing that it is possible to design AI systems which do not need a description or plan of their environments in order to engage with it nor do they need a centralized representation system. Following Brooks, the intelligence of these creatures was precisely non-representationalist in the sense that they worked on the basis of complex interrelations:

> Just as there is no central representation there is not even a central system. Each activity producing layer connects perception to action directly. It is only the observer of the Creature who imputes a central representation or central control. The Creature itself has none; it is a collection of competing behaviors. Out of the local chaos of their interactions there emerges, in the eye of an observer, a coherent pattern of behavior. [42] (pp. 148–149).

With regards to covariant information, however, REC is still in need of clarifying the nature of such informational relations. For instance, with their formalized definition, Hutto and Myin effectively exemplify information-as-covariance in relation to a non-cognitive phenomenon: the rings of a tree's trunk which covary with the tree's age. In this context, covariance is defined as follows: "s's being F 'carries information about' t's being H iff the occurrence of these states of affairs covary lawfully, or reliably enough" [13] (p. 66). But with regards to cognitive systems in the context of REC, there is no specific delineation of where and under what conditions such states of affairs exist. For instance, are both kinds of states of affairs extrinsic to the cognizer or might one or perhaps even both of them be internal (i.e., neural or organic) (for examples of such, see [43] (p. 67); [44] (p. 653)). Indeed, we do find evidence that Hutto and Myin allow the latter to be case such as in the following quote:

> The number of a tree's rings can covary with the age of the tree; however, this doesn't entail that the first state of affairs says or conveys anything true about the second, or vice versa. The same goes for states that happen to be inside agents and which reliably correspond with external states of affairs— these too, in and of themselves, don't "say" or "mean" anything just in virtue of instantiating covariance relations. [13] (p. 67).

Relatedly, as Hutto and Myin clarify, an appeal to covariance makes it possible to clarify the "correspondences to which the brain is sensitive" [41] (p. 238).

In this connection, the question is obviously whether we can consider basic cognition as involving a myriad of covariance relations in complex organisms? Here, a Code Biology framework can bring significant insights to the fore.

In their paper on signal-transduction codes, Marijuán and colleagues explore basic cellular communication in relation to E. coli K-12 [45]–which, interestingly, has also been a paradigm case for Autopoietic Enactivism [46]. Signal transduction is the process whereby the cell's membrane receptors transform "signals from the environment (first messengers) into internal signals (second messengers)" thus exhibiting how the cell adjusts or responds to changes in its environment [6] (p. 4). Initially, Marijuán and colleagues stress that in prokaryotic signaling processes, the molecular apparatuses involved can be either simple or complex. In the simple cases, we find the so-called one component systems which involve a "direct fusion of an input domain with an output domain, both put together in a single protein molecule" [45] (p. 30). Such simple systems entail that a signal molecule is immediately recognized by a cellular receptor which binds the external signal molecules [47] (p. 13). The cellular receptor acts as an adaptor: it mediates in ways that enable the messenger to function while allowing the cell to show sensitivity towards aspects of its environment (i.e., sensed external stimuli). Considered in REC-terms, we may say that such adaptors ensure the reliability in covariance processes in the sense of enabling particular signaling pathways whereby the cell can exhibit sensitivity towards the environment (in the specific form of sensed external stimuli) through the subsequent triggering of one or more functionally adequate intracellular responses. More specifically, given their sensitivity to the presence of vitamins, antioxidants etc. such basic component systems are vital to the cell's survival [45] (p. 30). In living organisms, covariance depends on multi-component systems that perform as adaptors. While these systems involve greater cell-internal complexity, they entail the mediacy which is the hallmark of the adaptor by ensuring the workings of intermediate mechanisms such as receptors (e.g., histidine kinases) and response regulators prior to the triggering of intracellular responses. Here, on the overall or general level of the cell it is a matter of which environmental stimuli effect (or covary with) the transcription of genes in the cell or as Barbieri puts it:

> The membrane receptors that implement signal transduction, furthermore, are molecular adaptors that create links between first and second messengers just as the transfer-RNAs create links between codons and amino acids. [6] (p. 4).

On the specific level, however, we see that different covariance relations function in conjunction in the sense that there is (1) the covariance between the external stimuli and the response sensor's (i.e., histidine kinase) transferring of a phosphoryl group and, following this, (2) a response regulator which in its activation of a transcription factor ensures covariance between the transferred phosphoryl group and the subsequent gene transcription [45] (p. 30). Yet, considering the fact that E. coli K-12 cells themselves are vastly complex in the sense of involving a myriad of more than one hundred of such simple and complex component systems [48], it becomes difficult, if not impossible, to delimit the singleness of the 'basic mind' on the basis of the covariant relations which constitute it. In fact, as Hofmeyr points out, given the interactional complexity (or, following Brooks, chaos) of E. coli K-12 cells, "the question of what the actual codes are becomes very difficult to answer" [47] (p. 13). Yet, what Code Biology here has to offer is a window into the complex mechanisms that enable adaptors to contribute to cellular responses which allow the cell to function as (or like) a 'basic mind'.

5 Towards a Synergistic Future?

One advantage for Code Biology is that it basically resonates with the enactivist assumption that agents bring forth aspects of their own environments. Given the fundamental arbitrariness of codes and, hence, the fact that life is different from pure chemical necessities and the spontaneity of non-living processes, code-making (and, hence, life as such) is seen as a constructive process: as artifact-making [9] (p. 6). In this connection, it is worth mentioning that a residue of what Cowley [35] terms the 'received view' nevertheless remains in Code Biology. For when it comes to information, there is a clear tendency to assume that codes involve the transfer of informational content which, effectively, is taken to be synonymous with Shannon-style information [13]. This links with the fact that Code Biology in the general sense has emerged out of semantic biology [8] and, thus, the view that information transfer can be considered as content-vehicle relations which unfold through linear exchanges (or transfers) in input-output systems. Whereas this could be emphasized as a decisive weakness of Code Biology that might bring it at odds with REC's take on basic mentality as being free of informational content, one could also see it as a positive aspect. If one decides to venture this route, two factors should be kept in mind: First, that the definition of a code is wide and not tied to a specific ontology which is underlined by the vagueness or generality pertaining to the so-called independent worlds which a given code is assumed to connect. Second, that in recent work, Paredes and colleagues have stressed the relevance of Bateson-style information to Code Biology and, hence, the fact that information can basically be described as 'a difference which makes a difference' [48]. The importance of this move can hardly be overstated in that it not only opens up for a radically different take on informational relations which does not presuppose Shannon's dualist ontology but also that it gives evidence of the fact that a Code Biology-framework can be compatible with different (or even: potentially contrasting) notions of information. In an enactivist context, and especially considered in relation to REC, we find that Code Biology might be the framework needed for not just overcoming REC's covariance-content distinction but also for unpacking the nature of qualitatively different informational relations more generally (cf. The example of the different levels of covariance in signal-transduction). Yet, one could argue that an appeal to 'information' is problematic in the context of Autopoietic Enactivism considering that Maturana and Varela explicitly contrast autopoiesis to notions such as 'coding', 'message', and 'information' [19] (p. 90). Yet, as remarked elsewhere [49], Maturana and Varela are here criticizing information as it was conceived by Shannon. Effectively, their criticism merely pertains to non-relationalist notions of information thus leaving it open for alternative conceptions of information to align with principles of autopoiesis.

Considering the two issues related to Code Biology's three worlds view (see Sect. 1.1), it is necessary to set aside this particular view. Otherwise, Code Biology (in the general sense) and Enactivism will be fundamentally at odds with each other, making future synergies impossible. One solution to this could be to reformulate some of the basic assumptions behind the three-world view, namely that different types of living beings (e.g., bacteria, horses, and humans) differ in terms of their semiotic complexity. At the same time, it is essential to refrain from considering these differences in terms of distinct ontological levels. Thus, to remain faithful to strong life-mind continuity, it is crucial to use the same basic notions for explaining semiosis (and, hence, basic cognitive phenomena) across different evolutionary scales.

While the generality of the ontology of codes–understood as the 'mapping between the objects of two independent worlds that is implemented by the objects of a third world called adaptors'–has shown to be quite useful due to its interpretative flexibility which opens up a space for interdisciplinary work, we also find that its very flexibility might become Code Biology's Achilles' heel. The reason for this is that it opens up for distinguishing between, on the one hand, recurrent codes and, on the other, relations which can be described as codified but which nevertheless might be so special, arbitrary or rare that they have no general constitutive relevance to biofunctions, neurofunctions or culture-related phenomena. This means that Code Biology has both an explanatory and a descriptive potential. Yet, for the sake of making it appealing in an enactivist context, it is the explanatory potential of Code Biology that would have to be developed further. In this connection, the co-development with enactivist positions can be potentially useful since the latter offer notions for systematically expanding code biological insights in relation to cognitive phenomena (e.g., 'teleology', 'basic minds', 'sense making', 'scaffolded cognition', 'sensorimotor contingencies'). Enactivist terminology can function as a theoretical anchoring point for developing Code Biology's theoretical framework which, so far, includes general notions such as 'codemakers', 'worlds', 'adaptors', 'meaning' etc. which could benefit from being made more specific in the general realm (i.e., beyond organic coding) including in the context of human-specific phenomena.

In this connection, it is clear that we must go beyond the Autopoietic Enactivist view on agency as being strongly autonomous. Indeed, Cowley's account on the relation between adaptors and codemakers suggests that human agency exhibits weak autonomy in the sense that cognition is informed by not just intrinsically derived values and norms, but also extrinsic ones. Moreover, his work also showcases how novelties are brought about as a result of the rule–and norm-followings which are pre-determined but nevertheless irreducible to what is inscribed in such rules and norms (and the codified relations determined by them). So, cognitive agency (at least in human behavior) is irreducible to the mere following of codes.

References

1. Barbieri, M.: Introduction to Code Biology. In: First International Conference in Code Biology. http://www.codebiology.org/conferences/pdf/conferens_papers/01.pdf. Accessed 07 Feb 2023
2. Crick, F.H.C., Barnett, L., Brenner, S., Watts-Tobin, R.J.: General nature of the genetic code for proteins. Nature **192**(4809), 1227–1232 (1961)
3. Turner, B.M.: Histone acetylation and an epigenetic code. BioEssays **22**(9), 836–845 (2000)
4. Strahl, B.D., Allis, C.D.: The language of covalent histone modifications. Nature **403**(6765) (2000)
5. Barbieri, M.: Code biology–a new science of life. Biosemiotics **5**, 411–437 (2012). https://doi.org/10.1007/s12304-012-9147-3
6. Barbieri, M. What is code biology? Biosystems **164**, 1–10 (2018). https://doi.org/10.1016/j.biosystems.2017.10.005
7. Barbieri, M.: Code Biology. A New Science of Life. Springer, Dordrecht (2015)

8. Barbieri, M.: The Organic Codes: An Introduction to Semantic Biology. Cambridge University Press, Cambridge (2003)
9. Barbieri, M.: Evolution of the genetic code: The ambiguity-reduction theory. Biosystems **185**, 104024 (2020)
10. Varela, F.J., Thompson, E., Rosch, E.: The Embodied Mind: Cognitive Science and Human Experience. MIT Press, Cambridge MA (1991)
11. Ward, D., Silverman, D., Villalobos, M.: Introduction: The Varieties of Enactivism. Topoi **36**, 365–375 (2017)
12. Barandiaran, X.E.: Autonomy and enactivism: towards a theory of sensorimotor autonomous agency. Topoi **36**, 409–430 (2017)
13. Hutto, D., Myin, E.: Radicalizing Enactivism: Basic Minds Without Content. MIT Press, Cambridge MA (2013)
14. De Jesus, P.: Autopoietic enactivism, phenomenology and the deep continuity between life and mind. Phenomenol. Cogn. Sci. **15**(2), 265–289 (2015). https://doi.org/10.1007/s11097-015-9414-2
15. Thompson, E.: Life and mind: from autopoiesis to neurophenomenology. A tribute to Francisco Varela. Phenomenol. Cogn. Sci. **3**, 381–398 (2004)
16. Di Paolo, E.: Autopoiesis, adaptivity, teleology, agency. Phenomenol. Cogn. Sci. **4**, 97–125 (2005)
17. Thompson, E.: Mind in Life: Biology, Phenomenology and the Sciences of Mind. Harvard University Press, Cambridge MA (2007)
18. Di Paolo, E., Thompson, E.: The enactive approach. In: Shapiro, L. (ed.) The Routledge Hand-Book of Embodied Cognition, pp. 68–78. Routledge Press, New York (2014)
19. Maturana, H.R., Varela, F.J.: Autopoiesis and Cognition: The Realization of the Living. D. Reidel Publishing Company, Dordrecth (1980)
20. Froese, T., Di Paolo, E.: The enactive approach: Theoretical sketches from cell to society. Pragmat. Cogn. **191**(1), 1–36 (2011)
21. Barbieri, M.: Codepoiesis–the deep logic of life. Biosemiotics **5**, 297–299 (2012). https://doi.org/10.1007/s12304-012-9162-4
22. Luisi, P.L.: Autopoiesis: a review and a reappraisal. Naturwissenschaften **90**, 49–59 (2003)
23. Cowley, S., Andersen, R.G.: Deflating autonomy: human interactivity in the emerging social world. Intellectica **63** (2015)
24. De Jesus, P.: Making sense of (autopoietic) enactive embodiment: a gentle appraisal. Phainomena **25**(98–99), 33–56 (2016)
25. Fanaya, P.F.: Autopoietic enactivism: action and representation re-examined under Peirce's light. Synthese **198**(Suppl 1), 461–483 (2021)
26. Gahrn-Andersen, R., Prinz, R.: Ensuring wholeness: using code biology to overcome the autonomy-heteronomy divide. Biosystems **26**, 104874 (2023)
27. De Jaegher, H., Di Paolo, E.: Participatory sense-making. Phenomenol. Cogn. Sci. **6**, 485–507 (2007)
28. Kirchhoff, M.D.: Autopoiesis, free energy, and the life–mind continuity thesis. Synthese **195**, 2519–2540 (2018). https://doi.org/10.1007/s11229-016-1100-6
29. Kirchhoff, M.: Predictive brains and embodied, enactive cognition: an introduction to the special issue. Synthese **195**, 2355–2366 (2018). https://doi.org/10.1007/s11229-017-1534-5
30. Allen, M., Friston, K.J.: From cognitivism to autopoiesis: towards a computational framework for the embodied mind. Synthese **195**, 2459–2482 (2018)
31. O'Regan, K., Noë, A.: A sensorimotor account of vision and visual consciousness. Behav. Brain Sci. **24**(5), 883–917 (2001)
32. Noë, A.: Concept pluralism, direct perception, and the fragility of presence. In: Metzinger, T., Windt, J.M. (eds) Open MIND **27**(T) (2015)

33. Clark, A., Toribio, J.: Doing without representing. Synthese **101**, 401–431 (1994)
34. Degenaar, J., O'Regan, J.K.: Sensorimotor theory and enactivism. Topoi **36**, 393–407 (2017). https://doi.org/10.1007/s11245-015-9338-z
35. Cowley, S.J.: Wide coding: Tetris, Morse and perhaps, language. Biosystems **185**, 104025 (2019)
36. Harris, R.: The Language Myth. Duckworth, London (1987)
37. Wilson, R.A.: Boundaries of the Mind: The Individual in the Fragile Sciences. Cambridge University Press, Cambridge (2004)
38. Farina, A.: Ecoacoustic codes and ecological complexity. Biosystems **164**, 147–154 (2018)
39. Hutto, D.D.: Knowing what? Radical versus conservative enactivism. Phenom. Cogn. Sci. **4**, 389–405 (2005). https://doi.org/10.1007/s11097-005-9001-z
40. Chemero, A.: Radical Embodied Cognitive Science. MIT Press, Cambridge MA (2009)
41. Hutto, D., Myin, E.: Evolving Enactivism: Basic Minds Meet Content. MIT Press, Cambridge MA (2017)
42. Brooks, R.A.: Intelligence without representation. Artif. Intell. **47**(1), 139–159 (1991)
43. Jacob, P.: What Minds Can Do: Intentionality in a Non-Intentional World. Cambridge University Press, New York (1997)
44. Abramova, K., Villalobos, M.: The apparent (Ur-)intentionality of living beings and the game of content. Philosophia **43**, 651–668 (2015). https://doi.org/10.1007/s11406-015-9620-8
45. Marijuán, P.C., Navarro, J., del Moral, R.: How prokaryotes 'encode' their environment: systemic tools for organizing the information flow. BioSystems **164**, 26–38 (2018). https://doi.org/10.1016/j.biosystems.2017.10.002
46. De Jesus, P.: Thinking through enactive agency: sense-making, bio-semiosis and the ontologies of organismic worlds. Phenom. Cogn. Sci. **17**(5), 861–887 (2018). https://doi.org/10.1007/s11097-018-9562-2
47. Hofmeyr, J.S.: The first Special Issue on code biology–A bird's-eye view. Biosystems **164**, 11–15 (2018)
48. Paredes, O., Morales, J.A., Mendizabal, A.P., Romo-Vázquez, R.: Metacode: one code to rule them all. Biosystems 104486 (2021). https://doi.org/10.1016/j.biosystems.2021.104486
49. Gahrn-Andersen, R.: Informational resilience in the human cognitive ecology. Entropy **25**, 1247 (2023). https://doi.org/10.3390/e25091247

Clinical Cognitive Sciences

Graham Pluck⬤ and Kris Ariyabuddhiphongs(✉)⬤

Faculty of Psychology, Chulalongkorn University, Bangkok, Thailand
`kris.ar@chula.ac.th`

Abstract. Clinical sciences involved with the mind and brain, including neurology, psychiatry, endocrinology and clinical psychology all frequently deal with cognitive symptoms, side effects, and risk factors. Consequently, there has long been some interaction between those clinical fields and traditional cognitive sciences, focused on computationalist and embodied approaches to understanding natural and machine cognition. Examples include the advances made in understanding the normal cognitive architecture made by studying its breakdown in disease, as well as the enhanced methods of defining and measuring cognitive disorders stemming from understanding the healthy state. Nevertheless, the fields currently fail to fully exploit the potential for mutual advancement. Here we explore the interactions between traditional clinical and cognitive sciences and highlighted strengths of the relationship, and areas that could benefit from greater multidisciplinary emphasis. We argue that original fields of cognitive science (philosophy, linguistics, computer science, anthropology, psychology and neuroscience) remain the core of the multidisciplinary cognitive sciences, but that they can all be applied fruitfully to clinical issues. We explore this in one sample disorder—voice hearing in schizophrenia, showing the potential for clinically applied cognitive sciences. It is our contention that greater achievement is possible, in both academic and applied fields dealing with cognition, if we can foster a mutually symbiotic relationship between the clinical and cognitive sciences.

Keywords: Neurology · Psychiatry · Cognitive disorders · Applied sciences · Hallucinations · Cognitive science

1 Introduction

The endeavor of cognitive science as an interdisciplinary science of the mind is often dated as beginning in earnest in 1956 [26]. Although in the nearly seven decades of progress since then, cognition has become an intensely studied topic, its obvious success is somewhat marred by frequent criticism of the disunity among the many fields that used the cognitive-computational metaphor (see, for example, the review by Nunez and colleagues in 2019 [32]). Although whether a unified cognitive science exists today remains a polemic point, there is no denying that the cognitive perspective has been very popular, particularly within

© The Author(s) 2024
A. Aldini (Ed.): SEFM 2023, LNCS 14568, pp. 130–148, 2024.
https://doi.org/10.1007/978-3-031-66021-4_9

psychology and neuroscience. Indeed, the majority of papers published in cognitive science journals nowadays are penned by psychologists. This fact is often lamented, suggesting a failure to balance the contributions from other core areas of the original cognitive science approach (i.e., linguistics, philosophy, anthropology and computer science). However, it has also been argued that despite the dominance of cognitive psychology, cognitive science nevertheless remains more interdisciplinary than either psychology or neuroscience (as judged by the authorship affiliation of study authors in leading journals [7]. Furthermore, Contreras Kallens and colleagues, who performed the audit of author affiliation, argue that "Cognitive science could 'grow into' its many disciplines by embracing new collaborators who inhabit our disciplinary silos, but who have not yet applied their trade to the core questions of our field." [7], p. 643. By this they point out that psychology, though dominant in many ways, is in reality very multidisciplinary, often with people doing linguistics, computation, anthropological work etc. within psychology departments as flags of convenience rather than any mark of being fundamentally psychologists. Though the wider point is simply that potential interdisciplinary links within sciences dealing with cognitive topics, defined broadly, are available and should be exploited more.

In this position paper, we particularly focus on how applied fields could interact more with academic cognitive science in a mutualistic symbiotic relationship. Of course, computer science, particularly artificial intelligence, has a strong applied aspect that contributes to cognitive science. However, we wish to focus on one particular applied topic, which we argue could lead to greater mutual benefits with cognitive science - clinical science. In recent years there has already been much research showing how cognitive biases and heuristic thinking influence clinical decision making. This input from cognitive science is welcome, however, there are many other connections that could be improved. Enhancing interdisciplinary or multidisciplinary work between cognitive and clinical sciences is a daunting task, given the very different approaches and needs of the two broad sciences, and their own heterogeneity. Nevertheless, mutually beneficial interactions, with cognitive studies learning from clinical cases, and the reverse, have a long history with proven examples of mutual benefit. Several examples of this are described in the sections below, so the endeavor can be fruitful. We simply encourage greater interaction. In Sect. 2 we make our basic position for why there exists potential for an applied clinical cognitive science. In Sect. 3 we describe ways in which cognitive and clinical sciences currently interact to mutual benefit, and ways that this could be further developed. In Sect. 4 we explore how clinical sciences fits within the traditional, interdisciplinary approach advocated by cognitive scientists. In the penultimate part, Sect. 5, we present an example of a clinical disorder that is fundamentally within the realm of cognitive science, and give examples of how different component disciplines of cognitive science contribute to understanding it. In Sect. 6, this paper concludes with some final observations and summation of the prospects for the nascent field of clinical cognitive sciences.

2   Why Clinical Cognitive Sciences?

Although definitions of what constitutes theory in cognitive sciences vary tremendously, one reasonable one is that "A cognitive theory is a description of mechanisms that explain observed mental phenomena" [46] p.239. The philosopher Paul Thagard argues for this definition because it is consistent with not just what happens in core cognitive sciences, but also with more peripheral fields that deal with cognition, such as clinical medicine. It is a rather obvious fact to mention that clinical impairments of brain function can reveal things about the human mind. That this is so well-known is reflected in the fact the earliest written description of the word 'brain', in the ancient Egyptian Edwin Smith papyrus, also contains the earliest ever description of impaired use of language consequent to brain damage [28]. Though, the extent to which clinical brain health and cognitive ability are intertwined is perhaps not so fully appreciated.

Disorders of the mind are often diagnosed and treated by clinical psychologists. As they are essentially applied psychologists, their interest spans the breadth of mental phenomena, albeit in the context of clinical disorders of the mind or brain. If one takes a more biomedical approach, focused specifically on the nervous system, the two principal medical sciences concerned with the brain are neurology (which deals with disorders defined organically, i.e., affecting the nervous system) and psychiatry (which deals with disorders defined by their impact on mental health- psychopathology). In both neurology [22] and psychiatry [44], the disorders observed usually involve cognitive processing impairments. As clinical psychologists deal with the same patient groups, the same can be said for clinical psychology. Even beyond the boundaries of the nervous system, the activity of many endocrine glands and the hormones that they release into the bloodstream influence brain functioning. Consequently, most disorders seen by clinical endocrinologists also involve alterations to cognitive processing [14]. Clearly, the majority of clinical disorders attended by clinical psychology and clinical neuroscience involve clinical signs, symptoms, side effects etc. that are essentially changes to cognitive ability. Furthermore, cognitive ability, in the form of intelligence, is recognized as a substantial protective factor against a wide range of disorders, a phenomenon known as cognitive reserve [45].

Adding to this, there has been a recent recognition that interoception (i.e., the sensation of signals from bodily organs outside of the nervous system) plays a much more important role in the mind than previously realized. In fact, it has been argued that a wide range of homeostatic mechanisms and bodily sensations are essential drivers of consciousness [10], implying that the whole body (the entire subject of clinical medicine) influences cognition. In addition, the discovery of mirror neurons has demonstrated that the same cells in the brain's premotor cortex that are involved in coordinating actions are also active during observation of actions [37], suggesting that the action system is also involved in perception of actions. It has been theorized that the mirror neuron system is involved in understanding the intentions of others as well as empathy. This latter point presents a strongly embodied perspective, as opposed to the more traditional computationalist approach in cognitive science. Here, we argue that whether or not one accepts an embodied or computationalist approach, it is undeniable that the physiological substrate of human cognition is in the body, and as such, bodily health influences cognitive processing. Following from this, clinical disorders will very frequently have cognitive correlates.

3 Mutual Symbiosis Between Cognitive and Clinical Sciences

In this section we first describe some of the ways in which clinical sciences contribute to understanding cognition. We then explore the reverse: how cognitive sciences contribute to clinical sciences. Here we highlight the mutual symbiosis of such collaborations. Nevertheless, it is likely that, given the differences in paradigms between the fields, and the fundamentally applied nature of medicine, compared to the oft purely academic nature of cognitive studies, that the interaction will be unbalanced. If anything, the benefits of collaboration so far have been reaped more by cognitive science than clinical sciences. This has been due to the revelatory nature of many disorders of the nervous system on cognitive functioning.

3.1 Clinical Disorders as Natural Experiments

As described above, as long as 5,000 years ago it had been noted by Egyptian scholars that aphasia can be caused by brain damage [28]. Relatively more recently, the issue of whether psychological traits, including cognitive abilities, show some level of modularity was famously addressed by the French neurologist Paul Broca in 1865 [2], when he revealed selective impairments of spoken language production in patients with damage to the left frontal lobe of the brain. Since then, damage to the brain has frequently been used as a natural experiment to elucidate the human cognitive architecture. This is known as the lesion-symptom mapping method. Although in one direction it is used to define functions of brain areas (i.e., cognitive neuroscience), it is similarly used to identify and define, at a strictly functional level, cognitive processes. When lesion-symptom mapping is used in this way to study the functional architecture of the mind, irrespective of physiological correlates, it is known as cognitive neuropsychology [6].

Several well-known observations in cognitive science were driven mainly by observations from cognitive neuropsychology, that is, cognitive impairments following brain damage. These include the distinction between procedural and declarative memory (often also known as implicit and explicit memory) first revealed by studies of patients with damage to the hippocampus producing dense amnesia who could nevertheless learn a mirror tracing task [27]. Another clear example being studies of neurological patients that indicated that visual perception for conscious recognition is relatively independent from visual perception for motor transformations [15].

The strength of these associations between brain impairment and cognitive impairment for elucidating the overall human cognitive system has been in the double dissociation method. This involves the comparison of patients with different cognitive problems, such that a patient can be demonstrated to be impaired on task x, but not task y, and another patient can be shown to have the opposite pattern of impairment and preservation of cognitive task performance. The logic behind the double dissociation is that cognitive processes x and y must be functionally independent if they can be impaired independently within the same overall cognitive system. The methodology allows for discounting of general explanations for the impairments, such as overall task performance being impaired, or global cognitive impairment, and supports the identification of cognitive modularity. A classic example of this has been the identification of patients with either preserved long-term memory (LTM) with impaired short-term memory (STM) and patients with the exact opposite pattern [48]. This double dissociation adds weight to the classic distinction between STM and LTM in Atkinson and Shiffrin's modal model of memory [1] and poses a serious challenge to cognitive theories which propose that there is only one declarative memory system that stores information [9].

Although lesion-deficit association studies still have some weaknesses, and alternative explanations for double dissociation which do not require modularity of function exist, for example from a neural network perspective [34], they undoubtedly have some role in cognitive sciences. The neuropsychologists Shallice and Cipolotti have listed several ways in which the traditional clinical method of studying brain-injured individuals has benefits over other cognitive sciences, including the potential for serendipitous discoveries, and identification of causal efficacy [42].

3.2   Clinical Disorders and Discovery of Cognitive Phenomena

We could also add that many cognitive neuropsychological disorders probably would not be predicted based on other methods in cognitive science. For example, stroke that causes brain damage very frequently produces a disorder in which patients neglect to attend to things to the left side of their body, or, less frequently, to the left side of individual objects. This syndrome, known as hemispatial neglect, appears to be fundamentally a disorder of the control of attention [8]. However, the reverse pattern (of attentional disorder to the right side of the body, or to objects) is much rarer. This suggests that multiple aspects of attention are not only fundamentally lateralized relative to the body of the observer, there is also a substantial lateralized imbalance, again operating left-right, relative to the observer. Further, related to this phenomenon, the clinical observation that stroke patients may show attentional neglect of either the left space (relative to their body) or to the left of the perceived objects, is now interpreted using the cognitive concepts of egocentric and allocentric spatial coding, respectively. This distinction too was first appreciated in clinical cases of brain damage [40].

Another example, also very common after stroke, is ideomotor apraxia—the inability to demonstrate learned actions such as tool use. Cognitive neuropsychological evaluations frequently find much worse performance for actions to verbal command, better performance for imitation, and best performance with the tool held in the hand [39]. This common observation in clinical neurosciences places constraints on cognitive models that aim to explain human tool use.

Evidence from neurological research also encourages the debate on the nature of cognitive architecture. For example, while apraxic patients may be unable to execute action plans related to use of objects, they retain knowledge about their identity, while other brain injured but non-apraxic patients show the reverse pattern [3]. This evidence from double dissociation suggests independent processing between declarative knowledge of object functions and motor-action plans for manipulation of objects. However, other neuropsychological studies show interconnectedness between cognitive abilities and motor systems. In an experiment involving Parkinson's disease patients, Nitiscò and colleagues demonstrated that motor simulation via reading and repeating hand-related action verbs could reduce upper limb tremor, suggesting that language processing of bodily action simulates the experience of action execution [30]. In the case of patients with amyotrophic lateral sclerosis, it has been found that, in addition to motor system impairment, patients had difficulty with action-related verbs [50]. This degradation of action-related knowledge was also associated with neurodegeneration in motor cortices of the brain. In sum, research in clinical disorders often informs cognitive theories and provides a substantive testing ground for hypotheses.

3.3 Selectivity of Cognitive Impairments and Facilitation of Functions

As described above, a multitude of changes to the human nervous system, and body in general, have implications for cognitive processing. Clinical disorders are clearly associated with deficits in cognition. If it were simply a case of illness resulting in some global lack of processing capacity there would be little to learn from their study, from a cognitive perspective. However, that is not the case. Disorders often manifest with relatively specific cognitive changes. The pattern of preservations and losses can therefore be highly informative about the overall cognitive architecture.

Furthermore, some clinical disorders are associated with better, not impaired cognition. For example, attempts at suicide and other acts of self-harm by patients with schizophrenia are more common in people with relatively good performance on word pronunciation tasks, compared to patients who do not self-harm [36]. Similarly, there are a range of clinical observations of enhanced cognitive performance on specific tasks after brain damage. These include better ability to detect deception from faces by people with aphasia and enhanced face detection in complex visual scenes in patients with visual agnosia (inability to recognize objects by sight), and recovery of attentional bias caused by a right hemisphere lesion, after a second lesion, this time to the left hemisphere [18]. Enhanced attention to detail, visuospatial activities, and perhaps even artistic ability is seen in some forms of dementia, and may even be useful in distinguishing between different forms of the disease [25]. Also, the observation that patients with schizophrenia develop better reading and spelling ability than education-matched control participants [19]. The many observations of enhanced cognition associated with clinical disorder undoubtedly have a wide-range of causes, which require deep understanding of how cognition is molded and enacted in the brain to explain them. As such, traditional approaches in clinical neuroscience, which implicitly use a disorders-cause-deficits paradigm, are generally insufficient [35]. What is needed is a greater appreciation, within clinical sciences, of the cognitive sciences.

3.4   Cognitive Science and Clinical Assessment

To take a task-based example, we can examine the Towers of Hanoi task. This involves three pegs and a set of disks of varying diameter. The task is to move the tower of disks from one peg to another, one disk at a time, with certain restrictions, such as never to place a larger disk on top of a smaller disk. This task, originally developed in mathematics, has attracted the attention of artificial intelligence, because of its multiple task versions and easily definable problem space [13]. It has also been used extensively in clinical neuropsychology, to measure cognitive planning ability in patients with neurological or psychiatric illness. In fact, it was selected for that purpose as a test that would particularly load on non-routine planning ability, and was hypothesized to be particularly sensitive to impairments of top-down cognitive control after damage to the frontal lobes of the brain [41].

The problem is that there are multiple ways to complete the Towers of Hanoi task. This has been demonstrated with various iterative, recursive and other algorithms in computer science [13]. Herbert Simon also demonstrated, from a cognitive science perspective, that humans who attempt the Towers of Hanoi have a wide range of strategies that they can use to successfully complete the task. Some strategies are transferable between different tasks, and some are not, and some require substantial use of working memory to represent sub-goal states, while others do not [43]. He emphasized that because so many different performance strategies and learning effects are involved, it is essential to examine performance on a subject-by-subject basis in order to estimate the cognitive mechanisms being used. Furthermore, cognitive science studies on choice of strategies in reasoning tasks have shown that they tend to vary across cultures [31], and even within cultures, reasoning strategy employed in tasks such as physics problems, varies by level of expertise on the material [21].

From a cognitive science perspective, Simon referred to these different ways of solving the same problems as *functional equivalence*. This parallelism between cognitive strategies used is also recognized in cognitive neuroscience. At the biological level, many cognitive processes show degeneracy, that is, the same behavioral outcome, such as word reading, or action imitation, can be achieved by different pathways within the brain [35]. Importantly, some of these degenerate pathways can become damaged, and others preserved, in the same patient, in which case no deficit will be observed. Thus, appreciation of functional equivalence / degeneracy is essential to recognizing clinical impairments of function.

Current clinical methods to evaluate top-down cognitive control use versions of the Towers of Hanoi task, but compare performance of individual clinical patients to average performance, regardless of strategies used. For example, the most-developed, commercially available test of top-down cognitive control for use in clinical practice is the Delis-Kaplan Executive Function System [11]. This includes a version of the Towers of Hanoi task, with scoring of performance primarily based on the number of moves made within time limits (fewer moves give higher scores). Other versions of the Towers of Hanoi use the same basic approach to scoring performance [41]. There are two main problems with this approach. The first problem is that patients are clinically evaluated for executive function impairments based on how well they perform the task (defined as lowest number of moves). Importantly, although they are told to use as few moves as possible, they are not told that they must complete it quickly. Hence, steady and careful planning could actually be penalized. Furthermore, those patients who promptly identify and apply one of the iterative strategies will be able to score highly, while patients who use different, but equally effective strategies will receive low scores, and perhaps be defined as cognitively impaired. The second problem is that individual performance is not evaluated in the subject-by-subject manner advised by Simon [43], rather, individual patients are compared to the average performance of a large group of healthy control participants. Thus, the control sample performance average will be calculated from task performance scores achieved using many of the different strategies that can validly complete the task.

Although many theoretical and experimental fields dealing with brain sciences, such as experimental neuropsychology, do often consider error types, and step-changes between trials that indicate changes in strategy (see e.g., [42]), this is often not the case in clinical neurosciences. Clinical sciences that deal with brain impairments often lack the sophistication of understanding of information processing present in the core cognitive sciences.

A final motive for the need for greater cooperation between clinical sciences and cognitive sciences comes from the cognitive models applied in clinical assessment. Often, the models used are outdated and misinterpreted. As an example, the Wechsler Memory Scale [49], widely used to define amnesic disorders in clinical practice, is overtly based around the modal model of memory proposed by the psychologists Atkinson and Shiffrin in 1968 [1]. Most cognitive scientists would see that as an outdated theory. Furthermore, the distinction between STM and LTM operationalized in that memory assessment, is that recall within 'several minutes' of stimuli exposure assesses STM, while recall after 25-30 minutes assesses LTM. That simplistic interpretation ignores most of what is known from cognitive sciences of the strategies for transfer from transient to long-term storage of information, and the intermediate stages of processing between them [4,5].

Clinical brain sciences could gain much from closer links to cognitive sciences. Related to this, is the emerging need for consistent cognitive ontologies. Many cognitive constructs, particularly in clinical sciences, are derived from common sense interpretations, or from general application of cognitive concepts, such as dysexecutive syndrome to describe a wide-range of impairments of cognitive and emotional control. However, there are now numerous attempts to harmonize the terms used across clinical and cognitive sciences [16]. Cognitive science is ideally placed to improve ontologies used in clinical practice and research regarding the brain.

4 The Place of Clinical Sciences within Cognitive Science(s)

Cognitive science is often conceived of as being composed of contributions from at least six different fields: philosophy, linguistics, anthropology, neuroscience, computer science, and psychology. A report in 1978 represented these as a hexagon, with each discipline at one of the vertices [26]. This is shown in Fig. 1.

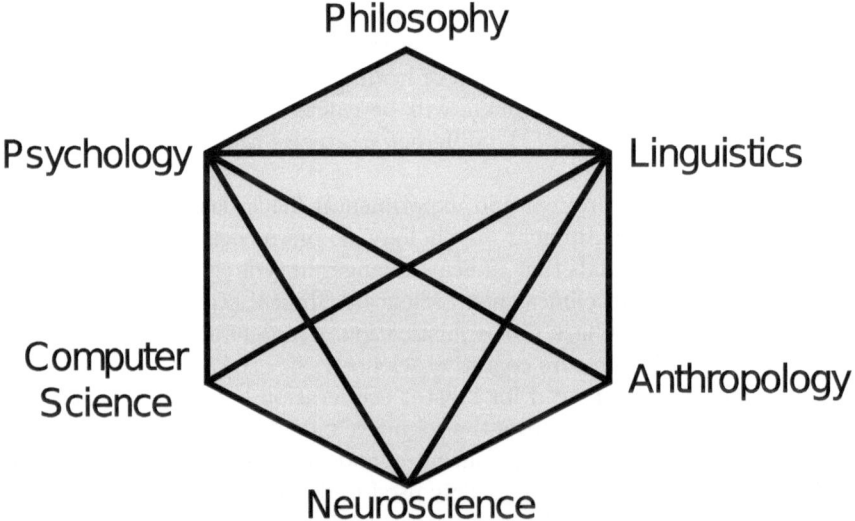

Fig. 1. The cognitive science hexagon, as envisaged in 1978, showing core cognitive fields and their viable interactions (shown as inter-connecting lines).

At that time of presentation in the late 1970's, only some of the fields were seen as having viable interdisciplinary subfields, for example, computer science was seen as interacting productively with psychology, neuroscience and linguistics, but not with philosophy or anthropology. This is shown by the 11 lines of the hexagon that connect them. It has since been argued that all of the interconnections have been achieved, and now, for example, it is reasonable to suggest that there is a philosophy of computer science, hence the newer version of the hexagon has all combinations of fields connected. The philosopher Paul Thagard, and George Miller [26,46], the originator of the hexagon, and arguably a founding father of cognitive science, concur. The revised version of the cognitive science hexagon (with 15 different interconnections) is shown in Fig. 2.

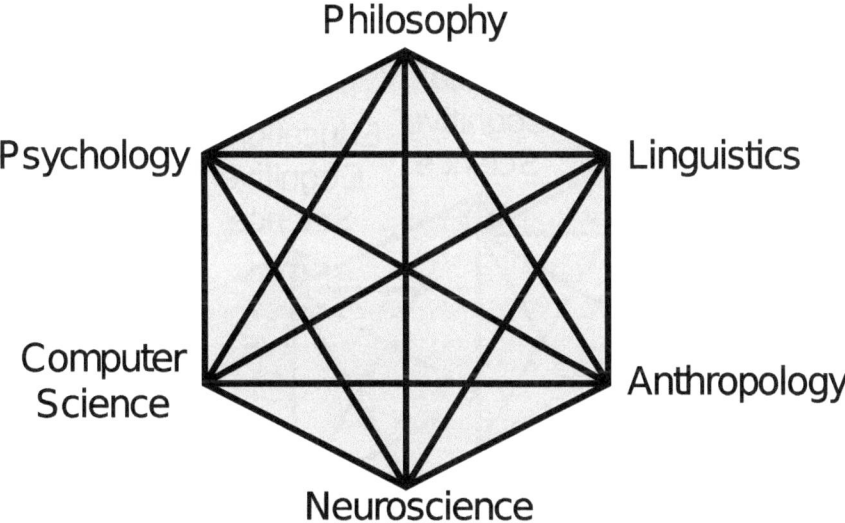

Fig. 2. The revised cognitive science hexagon, showing core cognitive fields with all connected by lines, indicating their potential for fruitful interactions.

Whether the six vertices can be considered to be working as an interdisciplinary cognitive science (singular), as originally envisaged, is debatable. The concept of interdisciplinarity suggests that research in each discipline is integrated and harmonized into a singular endeavor, with each field contributing more or less equally, but analysis of research output in cognitive science journals suggests that this is not the case [7,32]. Many researchers, including George Miller, now see the study of cognition as being more multidisciplinary [26], with disciplines focused on similar concepts, but having their own agendas, hence the increasing use of the term cognitive sciences (plural). As this term is said to indicate more multidisciplinarity [26,32], it seems appropriate for it to be used when applied to clinical matters.

One could easily argue that other endeavors could now be considered contributing fields to cognitive science, such as the emerging areas of cognitive design, cognitive history, and cognitive engineering. Indeed, some versions of the hexagram now include education (but shown as a heptagon). Although it is undeniable that cognitive approaches to understanding phenomena have expanded into many fields, it is debatable whether these additional fields have cognitive principles as core aspects. For this reason, we suggest that applied areas of cognitive science, such as education, should perhaps remain conceived of in terms of the original six vertices of the hexagon, but seen as applications of them. We demonstrate this idea graphically in Fig. 3. In the foreground we have the core cognitive sciences, and in the background their applications to diverse fields, such as educational, and most relevant to the current study—clinical cognitive sciences.

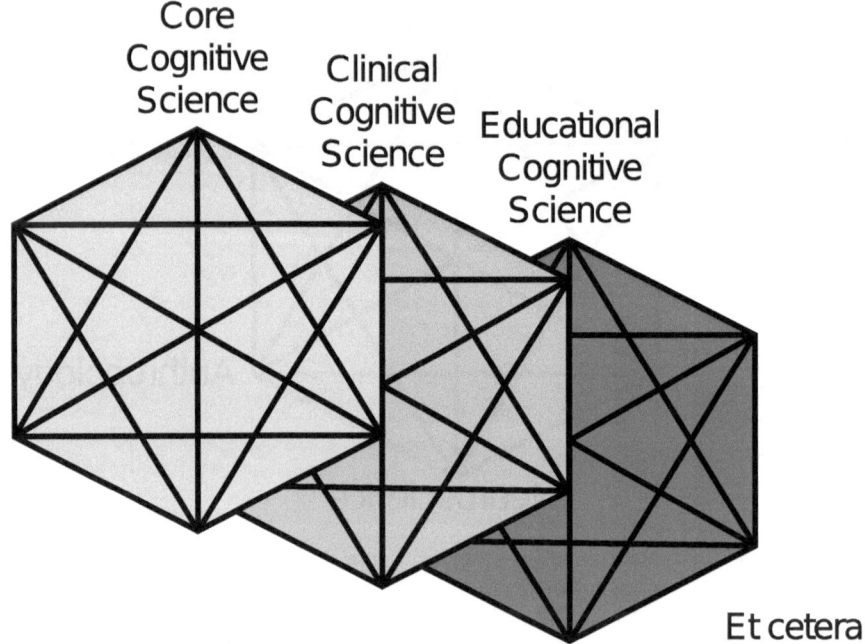

Fig. 3. A proposed scheme to think about the position of applied fields that interact substantially with the core cognitive sciences.

One could argue perhaps, that clinical disorders are already covered within the cognitive sciences under various investigations within neuroscience or psychology. But an important point is that fields such as those are primarily interested in the healthy state. There is a need for cognitive analyses that focus on the clinical issues. Furthermore, some issues that are important clinically, are of only tangential connection to psychology or neuroscience. A case in point is anosognosia, a state frequently seen in psychiatry and neurology, in which patients clearly have disease or disability, but because of their disorders are unable to recognize it [29]. For example, a dementia patient may be unaware that they have dementia, precisely because of their cognitive impairment, or a stroke patient with limb paralysis may be unaware of their paralysis. This is a clinical disorder that has substantial implications for understanding cognition, particularly from an embodied perspective. Anosognosia is important clinically due to the implications, such as compliance with treatment or care. In other words, clinical research could provide information related to construct validity of the cognitive theories. For example, the bottom-up approach to embodied cognition could be tested in the context of cognitive behavioral therapy (CBT) focusing on sensory-motor stimulation, while the top-down approach could be tested via CBT focusing on abstract mental representation of action knowledge [33].

5 Core Cognitive Science Contributions to a Sample Clinical Disorder

In the following section, we explore how the six original disciplines of cognitive science can individually contribute to understanding of clinical disorders. To do this we take as an illustration the hearing of voices within one's mind, that are not recognized as one's own. Such auditory hallucinations are a cardinal symptom of schizophrenia, but occur in many other medical states, and indeed, many people without any clinical disorder experience them too. Nevertheless, they can be very distressing, and as a symptom of schizophrenia, they are a core indicator of psychosis and are frequently treated with major tranquilizers. They are also fundamentally a cognitive phenomenon.

5.1 Philosophy and Cognition of Hearing Voices

That philosophy has a strong contribution to cognitive science is well established [46]. This often involves foundational issues such as the core theory of cognitive science. However, philosophy has also contributed substantially to understanding the phenomena of voice hearing in schizophrenia and other disorders. The clinical symptom of not recognizing voices within one's head as one's own has been particularly of interest to philosophers interested in understanding phenomenology. From this perspective, it has been argued that voices should not be considered as disorders at all, as neither a purely biological nor psychological explanation can account for their meaning. Instead, they can be thought of as embodied cognitive experiences, embedded in cultures which influence how voices are interpreted [47].

5.2 Cognitive Anthropology Approaches to Hearing Voices

The philosophical approach to voice hearing, drawing on the writings of phenomenological theorists, is broadly supported by cognitive research from anthropology. This field has examined how people describe their experiences of auditory hallucinations (which are usually voices). There appears to be a very wide range of experiences, not limited to clear voices. These include scratching, murmuring, whispering, with vague or clear contents, which can be psychologically located by the hearer either outside or inside their own heads. The anthropologist Tanya Luhrmann and colleagues have compared the experiences of auditory hallucinations in people with schizophrenia reported across cultures [23]. In California, they found that voice hearers tend to report diagnostic labels from psychiatry and refer to being 'crazy', and they uniformly disliked the voices that they heard. But this psychiatric vocabulary was very rare for patients in Accra, Ghana, or Chennai, India. Instead, the Ghanaian patients were likely to interpret the voices positively, while the Indian patients interpreted the voices as providing guidance. The anthropologists interpreted this by suggesting the people in the USA tend to interpret minds as being bounded, and thus unwanted voices must be pathological. In contrast the Ghanaian folk concept of mind is that it is porous- emotions seep into the world and can cause harm. Their interpretation of the voices was therefore supernatural, that voices and feelings were being controlled by God. In contrast, the Indian voice hearers interpreted the voices in terms of how older people provide guidance for people, often recognizing the voices as being kind or Hindu avatars.

To explain these differences, Luhrmann and colleagues suggested a form of 'social kindling' that alters how auditory hallucinations form. Drawing on cognitive psychology they argue that, due to cultural influences, people developing psychosis selectively attend to different aspects of their sensations. They cite evidence that this attentional focus then shapes how the auditory hallucination unfolds.

5.3   Cognitive Linguistics Approaches to Hearing Voices

The approach from psycholinguists concurs with that from anthropology, supporting the suggestion that how auditory experiences are interpreted influences their clinical presentation. Linguists particularly study the structure of language, when applied to auditory hallucinations this can include the content of the voices heard. However, the way that people with schizophrenia who hear voices describe their experiences is also important. One psycholinguistic study reported on the metaphors that patients use to describe their voices. This revealed that there was a remarkable consistency in the phenomenology of the voices, in that similar metaphors were used by all participants to indicate location and movement of the voices. However, the metaphors varied in terms of how distressed the patients were. Distressed patients described the voices using metaphors suggesting violence and lack of control (e.g., "it's like trying to fight with one hand behind your back" [12] p. 20. This association between distress and interpretation of heard voices is revealed through cognitive-psycholinguistic analysis.

5.4   Cognitive Psychology Approaches to Hearing Voices

Psychologists are primarily concerned with the normal functioning of the mind, and when applied to disorders, generally to explain the phenomenon in terms of breakdown of the normal system. One important contribution from psychology towards understanding auditory hallucinations has been the recognition that many people, not just people with clinical disorders, hear voices as auditory hallucinations. In fact, about one in ten healthy people will experience hearing voices in their lifetime, and consequently it is now considered as a phenomenon on a continuum from healthy to psychotic [24]. Drawing on this, it is argued that healthy people recognize that the experiences are generated internally, but people who experience the hallucinations as clinical symptoms, and often experience distress, may be failing to apply top-down executive control. This theory suggests that the voices are in fact normal perceptual processes related to auditory cognition. The reason that they may be misrepresented as being voices of strangers, being due to a failure of top-down inhibitory control, and this produces a strong attentional shift to the voice. This theory, based firmly in experimental psychology, is supported by experiments that use dichotic listening tasks [17]. These tasks present auditory stimuli, such as different syllables, to both ears simultaneously. The participant is asked to report what they hear. A right-ear advantage emerges in healthy individuals, thought to indicate the contralateral processing of auditory information in the left temporal lobe (specialized for phonology). Experimental evidence suggests that top-down executive control (hypothesized to the brain's prefrontal region) is limited in patients who hear voices due to functional disconnection from bottom-up processing in language centers (hypothesized to be in the brain's temporal region).

5.5 Cognitive Neuroscience Approaches to Hearing Voices

Studies on brain structure in people with schizophrenia who hear voices reach similar, but obviously more physiologically based conclusions. Reduced gray matter volume in auditory and language processing regions of the left temporal lobes of patients with schizophrenia is correlated with severity of their hallucinations, and when patients who hallucinate voices are compared to healthy individuals, the patients are found to have reduced gray matter volumes in the prefrontal cortex [38]. As gray matter volume indicates mainly neuronal cell bodies and dendrites (where synapses are present), it suggests reduced processing capability in those two regions, and disconnection between them, correspond to the language processing and inhibitory control modules suggested by experimental psychology.

5.6 Computer Science and Cognitive Modeling Approaches to Hearing Voices

Finally, research from computer science and artificial neural networks (ANNs) supports both the disconnection and inhibition approaches to understanding auditory hallucinations [20]. ANNs that are trained and then have disconnections induced, by extra pruning of connections to mimic the pruning of synapses, produce output suggestive of hallucinations. Similarly, relatively reduced levels of inhibitory connections in ANNs leads to confusion between bottom-up and top-down information, which could also be seen as hallucinatory and akin to hearing one's own internal voice as being not one's own, which is essentially the same as the psychological explanation for voice hearing in schizophrenia. These phenomena are currently receiving substantial attention within computer science, as it is becoming clear that 'hallucinations' are a common, and perhaps even inevitable feature of large language models such as ChatGPT.

5.7 Summary of the Cognitive Science of Auditory Hallucinations

Thus, the six fields that represent the core cognitive sciences have produced substantially overlapping ideas to help understand why people with schizophrenia frequently hear voices in their heads that they do not recognize as their own. Voice hearing like this would not usually be a topic in cognitive sciences, were it not that it is such a common clinical symptom. Furthermore, the analyses provided enrich the core of the cognitive endeavor. Voice hearing is but one example of cognitive disorders that can benefit from interaction with the field of cognitive sciences. While it may remain arguable whether the different fields work in an interdisciplinary way, it is clear that there is much to be gained from at least multidisciplinary application of the cognitive sciences to clinical problems.

6   Conclusions

The clinical sciences that deal with the mind and brain, including neurology, psychiatry, endocrinology and clinical psychology, already value the usefulness of applying cognitive principles to understanding disorders. And in the other direction, cognition has been frequently informed by clinical studies. However, greater integration would bring benefits to all sides. Nevertheless, the clinical and cognitive sciences have different basic paradigms, stemming from their strong emphases on biological and computational methods, respectively. This may, at times, present a barrier to cooperation. However, these need not be an insurmountable barrier, as the recent success in the fields of computational neuroscience and cognitive neuroscience have shown.

By identifying clinical cognitive sciences as an important applied parallel to the core academic cognitive sciences, we have attempted to bring greater attention to the mutually symbiotic relationship between clinical and cognitive sciences. In the spirit of applied technology, we would like to quote the industrialist Henry Ford: "Coming together is a beginning. Keeping together is progress. Working together is success". We feel that achievements so far are from cognitive and clinical sciences coming together, and keeping together. But much greater success is achievable from actively working together, in a mutually symbiotic relationship.

References

1. Atkinson, R.C., Shiffrin, R.M.: Human memory: a proposed system and its control processes, vol. 2, pp. 89–195. Academic Press, New York (1968). https://doi.org/10.1016/S0079-7421(08)60422-3
2. Broca, P.: Sur le siège de la faculté du langage articulé. Bull. Mem. Soc. Anthropol. Paris **6**, 377–393 (1865). https://doi.org/10.3406/bmsap.1865.9495

3. Buxbaum, L.J., Saffran, E.M.: Knowledge of object manipulation and object function: dissociations in apraxic and nonapraxic subjects. Brain Lang. **82**(2), 179–99 (2002). https://doi.org/10.1016/s0093-934x(02)00014-7

4. Cerone, A., Murzagaliyeva, D., Nabiyeva, N., Tyler, B., Pluck, G.: In silico simulations and analysis of human phonological working memory maintenance and learning mechanisms with behavior and reasoning description language (BRDL). In: Cerone, A., Autili, M., Bucaioni, A., Gomes, C., Graziani, P., Palmieri, M., Temperini, M., Venture, G. (eds.) Software Engineering and Formal Methods. SEFM 2021 Collocated Workshops, pp. 37–52. Springer International Publishing, Cham (2022)

5. Cerone, A., Pluck, G.: A formal model for emulating the generation of human knowledge in semantic memory. In: Bowles, J., Broccia, G., Nanni, M. (eds.) From Data to Models and Back, pp. 104–122. Springer International Publishing, Cham (2021)

6. Coltheart, M.: Lessons from cognitive neuropsychology for cognitive science: a reply to Patterson and Plaut (2009). Top. Cogn. Sci. **2**(1), 3–11 (2010). https://doi.org/10.1111/j.1756-8765.2009.01067.x

7. Contreras Kallens, P., Dale, R., Christiansen, M.H.: Quantifying interdisciplinarity in cognitive science and beyond. Top. Cogn. Sci. **14**(3), 634–645 (2022). https://doi.org/10.1111/tops.12609

8. Corbetta, M., Shulman, G.L.: Spatial neglect and attention networks. Annu. Rev. Neurosci. **34**, 569–99 (2011). https://doi.org/10.1146/annurev-neuro-061010-113731

9. Cowan, N.: What are the differences between long-term, short-term, and working memory? Prog. Brain Res. **169**, 323–38 (2008). https://doi.org/10.1016/S0079-6123(07)00020-9

10. Damasio, A., Damasio, H.: Feelings are the source of consciousness. Neural Comput. **35**(3), 277–286 (2023). https://doi.org/10.1162/neco_a_01521

11. Delis, D., Kaplan, E., Kramer, J.: Delis-Kaplan Executive Function System Technical Manual. The Psychological Corporation, San Antonio, TX (2001)

12. Demjén, Z., Marszalek, A., Semino, E., Varese, F.: Metaphor framing and distress in lived-experience accounts of voice-hearing. Psychosis **11**(1), 16–27 (2019). https://doi.org/10.1080/17522439.2018.1563626

13. Er, M.: A representation approach to the Tower of Hanoi problem. Comput. J. **25**(4), 442–447 (1982). https://doi.org/10.1093/comjnl/25.4.442

14. Erlanger, D.M., Tremont, G., Davis, J.D.: The neuropsychology of endocrine disorders. In: The Handbook of Clinical Neuropsychology, 2nd edn. Oxford University Press (2010). https://doi.org/10.1093/acprof:oso/9780199234110.003.30

15. Goodale, M.A., Milner, A.D.: Separate visual pathways for perception and action. Trends Neurosci. **15**(1), 20–5 (1992). https://doi.org/10.1016/0166-2236(92)90344-8

16. Hastings, J., Frishkoff, G.A., Smith, B., Jensen, M., Poldrack, R.A., Lomax, J., Bandrowski, A., Imam, F., Turner, J.A., Martone, M.E.: Interdisciplinary perspectives on the development, integration, and application of cognitive ontologies. Front. Neuroinform. **8**, 62 (2014). https://doi.org/10.3389/fninf.2014.00062

17. Hugdahl, K.: "Hearing voices": auditory hallucinations as failure of top-down control of bottom-up perceptual processes. Scand. J. Psychol. **50**(6), 553–60 (2009). https://doi.org/10.1111/j.1467-9450.2009.00775.x

18. Kapur, N., Cole, J., Manly, T.: The paradoxical brain. Psychologist **26**(2), 102–105 (2013)

19. Kremen, W.S., Seidman, L.J., Faraone, S.V., Pepple, J.R., Lyons, M.J., Tsuang, M.T.: The "3 Rs" and neuropsychological function in schizophrenia: an empirical test of the matching fallacy. Neuropsychology **10**(1), 22–31 (1996). https://doi.org/10.1037/0894-4105.10.1.22

20. Lanillos, P., Oliva, D., Philippsen, A., Yamashita, Y., Nagai, Y., Cheng, G.: A review on neural network models of schizophrenia and autism spectrum disorder. Neural Netw. **122**, 338–363 (2020). https://doi.org/10.1016/j.neunet.2019.10.014

21. Larkin, J.H., McDermott, J., Simon, D.P., Simon, H.A.: Models of competence in solving physics problems. Cogn. Sci. **4**(4), 317–345 (1980). https://doi.org/10.1016/S0364-0213(80)80008-5

22. Larner, A.J.: Neuropsychological Neurology: The Neurocognitive Impairments of Neurological Disorders. Cambridge University Press, Cambridge, UK (2013). https://doi.org/10.1017/CBO9781139176095

23. Luhrmann, T.M., Padmavati, R., Tharoor, H., Osei, A.: Hearing voices in different cultures: a social kindling hypothesis. Top. Cogn. Sci. **7**(4), 646–663 (2015). https://doi.org/10.1111/tops.12158

24. Maijer, K., Begemann, M.J.H., Palmen, S., Leucht, S., Sommer, I.E.C.: Auditory hallucinations across the lifespan: a systematic review and meta-analysis. Psychol. Med. **48**(6), 879–888 (2018). https://doi.org/10.1017/S0033291717002367

25. Midorikawa, A., Kumfor, F., Leyton, C.E., Foxe, D., Landin-Romero, R., Hodges, J.R., Piguet, O.: Characterisation of "positive" behaviours in primary progressive aphasias. Dement. Geriatr. Cogn. Disord. **44**(3–4), 119–128 (2017). https://doi.org/10.1159/000478852

26. Miller, G.A.: The cognitive revolution: a historical perspective. Trends Cogn. Sci. **7**(3), 141–144 (2003). https://doi.org/10.1016/s1364-6613(03)00029-9

27. Milner, B.: Les troubles de la memoire accompagnant des lesions hippocampiques bilaterales (1962). https://doi.org/10.1016/0028-3932(65)90029-1

28. Minagar, A., Ragheb, J., Kelley, R.E.: The Edwin Smith surgical papyrus: description and analysis of the earliest case of aphasia. J. Med. Biogr. **11**(2), 114–7 (2003). https://doi.org/10.1177/096777200301100214

29. Mograbi, D.C., Morris, R.G.: Anosognosia. Cortex **103**, 385–386 (2018). https://doi.org/10.1016/j.cortex.2018.04.001

30. Nistico, R., Cerasa, A., Olivadese, G., Dalla Volta, R., Crasa, M., Vasta, R., Gramigna, V., Vescio, B., Barbagallo, G., Chiriaco, C., Quattrone, A., Salsone, M., Novellino, F., Arabia, G., Nicoletti, G., Morelli, M., Quattrone, A.: The embodiment of language in tremor-dominant Parkinson's disease patients. Brain Cogn. **135**, 103586 (2019). https://doi.org/10.1016/j.bandc.2019.103586

31. Norenzayan, A., Smith, E.E., Kim, B.J., Nisbett, R.E.: Cultural preferences for formal versus intuitive reasoning. Cogn. Sci. **26**(5), 653–684 (2002). https://doi.org/10.1207/s15516709cog2605_4

32. Nunez, R., Allen, M., Gao, R., Miller Rigoli, C., Relaford-Doyle, J., Semenuks, A.: What happened to cognitive science? Nat. Hum. Behav. **3**(8), 782–791 (2019). https://doi.org/10.1038/s41562-019-0626-2

33. Pietrzak, T., Lohr, C., Jahn, B., Hauke, G.: Embodied cognition and the direct induction of affect as a compliment to cognitive behavioural therapy. Behav. Sci. **8**(3), 29 (2018). https://doi.org/10.3390/bs8030029

34. Plaut, D.C.: Double dissociation without modularity: evidence from connectionist neuropsychology. J. Clin. Exp. Neuropsychol. **17**(2), 291–321 (1995). https://doi.org/10.1080/01688639508405124

35. Pluck, G.: The misguided veneration of averageness in clinical neuroscience: a call to value diversity over typicality. Brain Sci. **13**(6), 860 (2023). https://doi.org/10. 3390/brainsci13060860

36. Pluck, G., Lekka, N.P., Sarkar, S., Lee, K.H., Bath, P.A., Sharif, O., Woodruff, P.W.: Clinical and neuropsychological aspects of non-fatal self-harm in schizophrenia. Eur. Psychiatry **28**(6), 344–8 (2013). https://doi.org/10.1016/j.eurpsy.2012. 08.003

37. Rizzolatti, G., Craighero, L.: The mirror-neuron system. Annu. Rev. Neurosci. **27**, 169–92 (2004). https://doi.org/10.1146/annurev.neuro.27.070203.144230

38. Romeo, Z., Spironelli, C.: Hearing voices in the head: two meta-analyses on structural correlates of auditory hallucinations in schizophrenia. Neuroimage Clin. **36**, 103241 (2022). https://doi.org/10.1016/j.nicl.2022.103241

39. Schnider, A., Hanlon, R.E., Alexander, D.N., Benson, D.F.: Ideomotor apraxia: behavioral dimensions and neuroanatomical basis. Brain Lang. **58**(1), 125–36 (1997). https://doi.org/10.1006/brln.1997.1770

40. Semmes, J., Weinstein, S., Ghent, L., Teuber, H.L.: Correlates of impaired orientation in personal and extrapersonal space. Brain **86**, 747–72 (1963). https://doi. org/10.1093/brain/86.4.747

41. Shallice, T.: Specific impairments of planning. Philos. Trans. R. Soc. Lond. B Biol. Sci. **298**(1089), 199–209 (1982). https://doi.org/10.1098/rstb.1982.0082

42. Shallice, T., Cipolotti, L.: The prefrontal cortex and neurological impairments of active thought. Annu. Rev. Psychol. **69**, 157–180 (2018). https://doi.org/10.1146/ annurev-psych-010416-044123

43. Simon, H.: The functional equivalence of problem solving skills. Cogn. Psychol. **7**(2), 268–288 (1975). https://doi.org/10.1016/0010-0285(75)90012-2

44. Snyder, H.R., Miyake, A., Hankin, B.L.: Advancing understanding of executive function impairments and psychopathology: bridging the gap between clinical and cognitive approaches. Front. Psychol. **6**, 328 (2015). https://doi.org/10.3389/fpsyg. 2015.00328

45. Stern, Y.: Cognitive reserve. Neuropsychologia **47**(10), 2015–28 (2009). https:// doi.org/10.1016/j.neuropsychologia.2009.03.004

46. Thagard, P.: Why cognitive science needs philosophy and vice versa. Top. Cogn. Sci. **1**, 237–254 (2009). https://doi.org/10.1111/j.1756-8765.2009.01016.x

47. Thomas, P., Bracken, P., Leudar, I.: Hearing voices: a phenomenological-hermeneutic approach. Cogn. Neuropsychiatry **9**(1–2), 13–23 (2004). https://doi. org/10.1080/13546800344000138

48. Warrington, E.K.: The double dissociation of short-and long-term memory. In: Cermak, L. (ed.) Human Memory and Amnesia, pp. 61–76. Psychology Press, Hove, UK (2014)

49. Wechsler, D.: Wechsler Memory Scale Third Edition Abbreviated Manual. The Psychological Corporation, San Antontio, TX (1997)

50. York, C., Olm, C., Boller, A., McCluskey, L., Elman, L., Haley, J., Seltzer, E., Chahine, L., Woo, J., Rascovsky, K., McMillan, C., Grossman, M.: Action verb comprehension in amyotrophic lateral sclerosis and Parkinson's disease. J. Neurol. **261**(6), 1073–9 (2014). https://doi.org/10.1007/s00415-014-7314-y

OpenCERT 2023—11th International Workshop on Open Community Approaches to Education, Research and Technology

OpenCERT 2023 Organizers' Message

The 11th International Workshop on Open Community Approaches to Education, Research and Technology (OpenCERT 2023) aimed to promote the use of Open Community approaches in Education and Research while also exploiting them to achieve wide diffusion and proper assessment of new, innovative Technology.

The workshop received four full paper submissions, which were reviewed for quality, correctness, originality and relevance. Each submission was posted on GitHub and reviewed by at least three Program Committee members. This first phase of the review process was carried out as an interactive, open discussion between the authors and the reviewers. Authors then had the chance to revise their works and submit the revised versions. A final review process, based on the revised papers, was carried out by the same reviewers of the first phase using the EasyChair system and was concluded by a closed discussion among the PC members. One contribution was presented at the workshop and is included in this volume: "Facets of Openness in a Serious Game: Opening up Format, Content, Software and Hardware" by Donatella Persico and Francesca Pozzi.

The workshop programme also featured one keynote talk: "I have a dream: AI to help people with Dyslexia in Education" by Filippo Sciarrone, Universitas Mercatorum, Italy.

We would like to thank the keynote speaker and the presenter of the contributed paper for their very stimulating talks. We are also grateful to the Program Committee members, for their enthusiasm and effort in actively participating in the open review process and for their extensive work in providing constructive feedback and collaborating with the authors during the revision process, and the authors, for their valuable contributions and for wisely exploiting the collaboration with the reviewers to improve their works. Finally, we would like to thank all workshop attendees for their active participation in discussions and for the feedback they provided to the speakers.

Antonio Cerone
Marco Temperini
June 2024

OpenCERT 2023 Organization

Program Committee Chairs

Antonio Cerone   Nazarbayev University, Kazakhstan
Marco Temperini Sapienza University Rome, Italy

Steering Committee

Luís Soares Barbosa University of Minho and UNU-EGOV, Portugal

Peter T. Breuer   Hecusys LLC, USA
Antonio Cerone   Nazarbayev University, Kazakhstan

Program Committee

Facets of Openness in a Serious Game: Opening up Format, Content, Software and Hardware

Donatella Persico(✉) ⓘ and Francesca Pozzi ⓘ

Consiglio Nazionale delle Ricerche, Istituto per le Tecnologie Didattiche, via de Marini 6, 16149 Genova, Italy
{donatella.persico,francesca.pozzi}@itd.cnr.it

Abstract. This paper advocates the claim that open licenses and open-source software are not enough to overcome the barriers to adoption of technologically innovative Open Educational Resources for educational institutions like schools. The paper analyses the case of the 4Ts game, a game designed to support the development of teachers' learning design skills. This case study exemplifies ways of dealing with four different facets of openness, namely format, content, software and hardware facets. The paper concludes that OERs for schools need to be flexible in terms of format and easy to amend in terms of content. Software should incorporate built-in features for personalization and localization that do not require coding skills. Hardware should be cheap and/or commonly found in schools.

Keywords: Open Educational Resources (OERs) · Teacher training · Game based learning · OER adoption · Serious games · Board games

1 Introduction

Investigation of openness in educational research can be dated back to the beginning of the '90s, when the first database of so-called "Units of Learning Material" was conceived and designed in the context of the ESM-BASE European project [1]. At the time, the central idea was to build repositories that would make it easy for instructional designers and teachers to access and retrieve self-consistent chunks of reusable educational material so they could reuse and repurpose these for different contexts. The need addressed was making multimedia development easier, increase quality and, at the same time, widen the target audience.

However, it was only during the subsequent decade that the concept Open Educational Resources (OER) attracted widespread attention after the worldwide-web gave a global dimension to the words "accessibility" and "reuse" [2], and many curated repositories of OERs, such as MERLOT,[1] were created. The important features of these repositories is that (i) they allow retrieval of material by discipline, language, school level and other features and (ii) the open license of that material specifies the extent to which anyone

---

[1] https://www.merlot.org/.

can reuse, revise, remix and redistribute the resources (the "4Rs" of OERs) [3]. In addition, in most cases the quality of the resources on offer is assured to some degree by the (curating) institution in charge of the repository. Hence, teachers do not need to reinvent themselves as multimedia developers, and students themselves can search and take advantage of material that satisfies their learning needs, provided they are self-regulated enough.

Research on OERs has shown that students using OER perform as well as, or better, than those using traditional materials [4–6]. In particular, Tlili and colleagues' meta-analysis [5] found that OER use has a positive significant (yet small) effect on learning outcomes, moderated by several variables, including subject, level of education and others. In addition, OERs (compared to traditional material) can reduce educational costs for students and institutions [7]. Finally, OERs have the potential to foster equity in education by providing access to learning material regardless of geographical location or socioeconomic status. However, research suggests that to ensure relevance and accessibility on a global scale, there is a need to consider diverse cultural and linguistic contexts in OER development and dissemination [2].

Research has also addressed concerns about the quality and sustainability of OER by proposing quality assurance mechanisms for OER repositories and exploring models for sustaining OER initiatives over time [8, 9].

In spite of these efforts, OER adoption is not as widespread as we could expect, and much of it takes place "under the radar" [10, 11]. Hence, researchers have explored factors influencing faculty adoption of OER and perceptions towards these resources [12, 13], while international organizations like UNESCO have issued recommendations [14] concerning national policies that could foster OER adoption. In a similar vein, the European Commission (EC) is also promoting the openness of EC-funded projects results by encouraging, and in some cases requiring, that their outputs be issued with an open copyright license, in line with the open science and principles. This is also true of any educational material produced within the framework of the Erasmus + programme[2], whatever its format. In spite of these important policies, [15] commentary "calls for a wider discussion to remove a number of barriers to mainstreaming OER in teaching and learning and argues for a rethinking of the idea of 'open' to make it more inclusive by redefining the concept" ([15], p. 369). In line with this call, in this paper we argue that we should be as flexible as possible when developing OERs to anticipate problems that might prevent uptake.

Against the above-described backdrop, this paper discusses a number of different nuances that the term *open* can take when the OER is a complex innovative resource comprising tangible, software, and hardware components intended for use in schools. The discussion will revolve around the '4Ts Game' [16–18], which was developed to support groups of teachers while designing collaborative teaching/learning activities for their students. We will maintain that, when schools are the target, extra caution is needed to make sure teachers will be able to use a resource, even before they can reuse it. Hence, in the following we will first describe the game and then illustrate the different facets of openness that were dealt with to make sure it can be used in different educational contexts. Finally, we draw some conclusions concerning the importance of making OERs

---

[2] https://erasmus-plus.ec.europa.eu/programme-giude/part-a.

as flexible as possible so as to minimize the effort required of teachers to use (and reuse) them.

2   The 4Ts Game

The '4Ts Game'[3] has been under development since 2015 and tested (via a user-centered design approach) with different international cohorts of teachers within two Erasmus + projects: PLEIADE[4] and SuperRED[5]. Both of these projects sought to develop the competence of European teachers in the design of collaborative learning activities for students, so the game was at the core of the respective teacher training interventions.

The game is based on the 4Ts theoretical model [19], which frames the design of collaborative learning activities as a complex decision-making process that encompasses four variables: the 'Task' (what students are asked to do); the 'Team' (how students will be grouped to perform the task together); the 'Time' (phases and schedule for accomplishing the task); and the 'Technology' (the technological tools and resources needed to do it).

According to the 4Ts model, designing a collaborative activity means making decisions concerning these four variables in order to achieve the learning aims in the educational context at hand. As choices concerning each of these variables have an influence on the others, the design process is iterative and may require several rounds to fine tune and optimize the design. According to the literature on collaborative learning, designers' decisions can be made in accordance with well-established techniques [20]), that is, content-independent patterns that help to structure collaborative learning activities. To clarify the concept, Fig. 1 shows an example of a technique, namely peer review, and a schematic representation of how it can be implemented using the 4Ts model.

The 4Ts game is a board game engaging groups of teachers in an interactive, reflective decision-making process centered around the four variables (the 4Ts) of the model and their interrelations. While the board represents the timeline of the activity being designed (that is the time variable), with each column representing one week, the techniques and the other three variables of the 4Ts model are represented by four different decks of cards (blue cards for techniques, red for tasks, yellow for teams, green for technology). The different cards contain indications on how they can be combined on the board. The first version of the game [18] is paper-based and can be played by groups of teachers standing around a table where the board lays. When playing the game, teachers make their decisions about the four variables by reading the cards, discussing what card combination is best suited for the activity at hand, choosing the agreed cards from the four decks, and positioning them on the board, as shown in Fig. 2. In this version of the game, a tutor is needed to assist teachers during gameplay and to provide feedback about their choices.

This version of the game was tested and fine-tuned in the course of a number of real-life experiments [16]. This phase paved the way for development of digital and hybrid versions of the game that reduces the need of tutor assistance while making sure the

---

[3] https://sites.itd.cnr.it/4TsGame/.

[4] https://pleiade-project.eu/.

[5] https://www.superred.eu/.

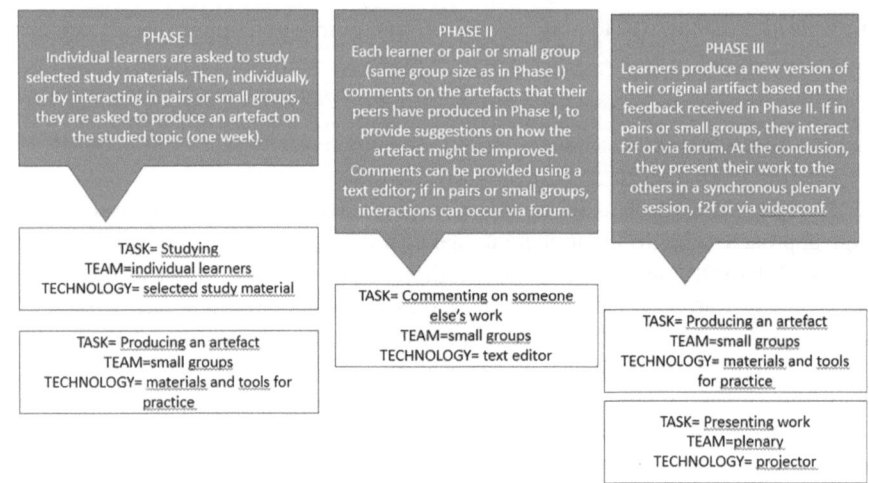

Fig. 1. How the peer review can be represented according to the 4Ts model.

Fig. 2. Teachers playing with the paper version of the 4Ts game.

playing teams receive instant feedback on their game moves. These two game versions also allow players to save a persistent configuration of the board, so that gameplay can be easily paused and resumed, an affordance not featured in the paper version.

In the digital version of the game, the board is reproduced on an Interactive White Board (IWB). The teacher group standing by the IWB (Fig. 3) can choose the cards they wish to play from virtual decks of cards displayed on the screen(Fig. 4). Whenever a card is played that is not compliant with the board configuration the game software provides feedback (Fig. 5) and, upon request, offers suggestions about what cards can be played. The digital game can also indicate whether the technique representation is complete (Fig. 6).

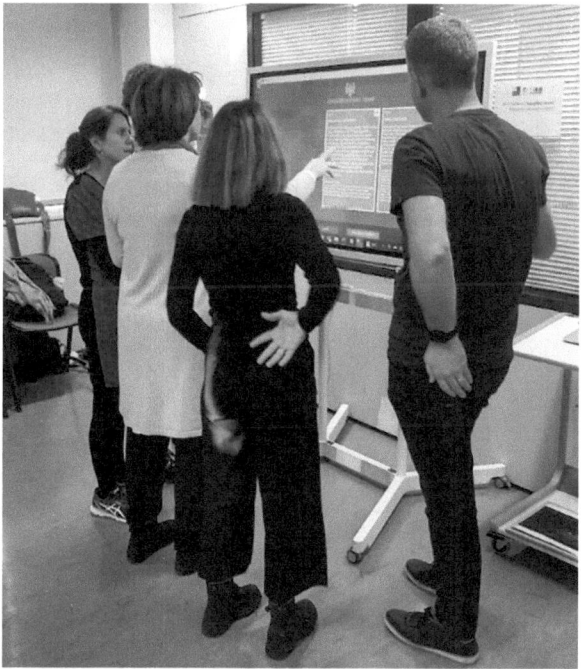

Fig. 3. Teachers playing with the digital version of the 4Ts game.

Finally, the hybrid version of the game (Fig. 7) allows teachers to play using the physical board and decks of cards, just like in the paper version. Here, however, players receive automatic feedback as in the digital version. To achieve this, both the board and the cards have ArUco Markers QR codes and a camera hanging above the board detects the cards as soon as the teachers place them on the board. In this way, the software component of the game can process players' moves and activate the same type of feedback as in the digital version. This is displayed on a PC, positioned to the side of the board, where the board configuration is replicated. A more detailed description of the digital and hybrid versions of the game is provided in [16]. A rationale for the need addressed by the game can be found in [22].

The digital and hybrid games can be played at three progressive levels of difficulty. At each new level, the degree of freedom teachers are afforded in decision making

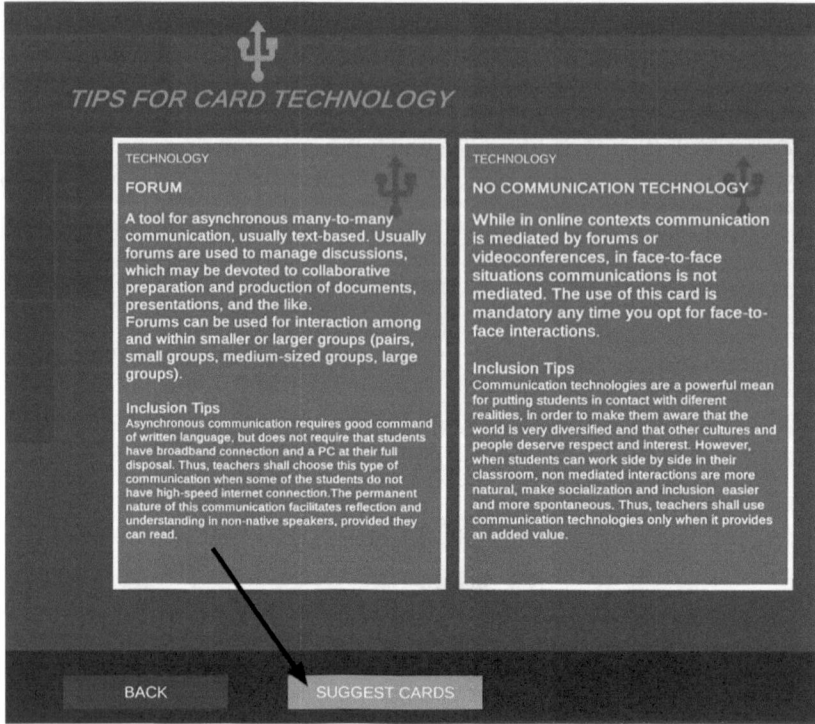

Fig. 4. Choosing from a deck of cards in the 4Ts game. After clicking the 'suggest cards' button, the game displays only those cards that are compliant with the state of the board.

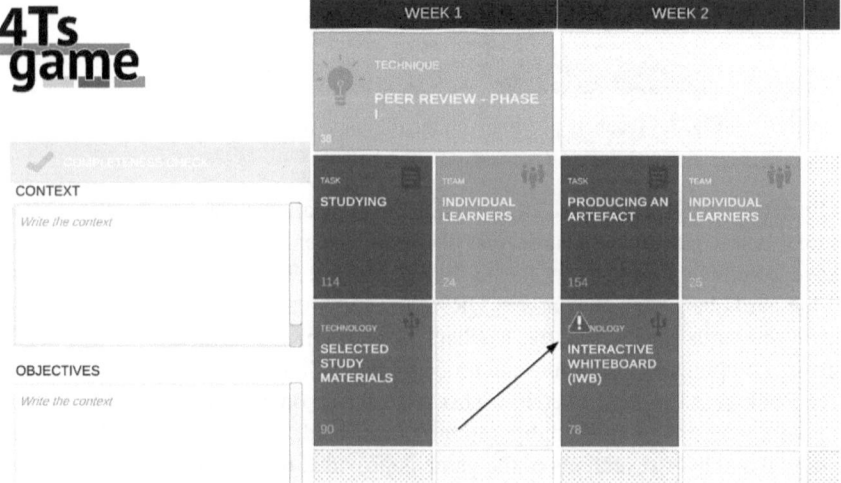

Fig. 5. The red triangle indicated by the arrow shows when the wrong card has been played.

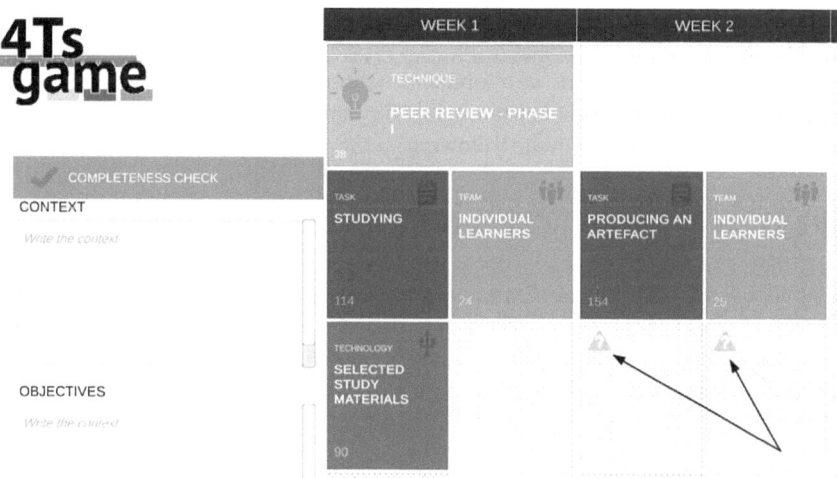

Fig. 6. The completeness check indicates the next board slots to be filled in.

Fig. 7. Teachers playing the hybrid version of the 4Ts game.

progressively increases, while the guidance provided decreases. Further details of the digital game functionalities and architecture are provided in [17].

As mentioned above, game development has been carried out through a user-centered approach, entailing play sessions conducted during several training interventions with different teacher cohorts. So far, the whole process has involved, more than 150 teachers from six different European countries (Italy, Bulgaria, Cyprus, Greece, Spain and Belgium). These real-world experiments with the game allowed us to assess game acceptance, ascertain the effectiveness of the approach, and compare both of these aspects across the different versions of the game. These qualitative and quantitative data are provided and discussed in previous publications concerning the game [16, 21] and have informed game design choices up to now. However, the focus of this paper is not on game acceptance or effectiveness, but rather on ways to facilitate its uptake by teachers working in different contexts. Evidence of success in this regard can only be collected in the long term, well after the end of the two projects (PLEIADE and SuperRED) that provided the conditions for its development. At the time of writing, we can only say that the teachers involved found customizing the game to their local context to be rather quick and easy, as discussed in the following.

3   Facets of Openness of the 4Ts Game

As mentioned in the introduction, this paper proposes the authors' reflections concerning a number of facets of the term *open*, as interpreted during the development of the 4Ts game. Clearly, this particular OER is a complex one, not just because it is a game, but because it is a complex and cutting-edge piece of technology intended to be used in teacher professional development, possibly in schools or in teacher training institutions, with or without the support of a teacher educator.

As [15] posits, the original thinking behind OERs was "to create universally available educational resources that can improve the quality of teaching and learning" (p.369). The expression "universally available" also means "universally usable", where the concept of "usability" necessarily relates to typical contexts of use and adoption by typical users. If OER features clash with the needs and affordances of the operational contexts of prospective users (in our case, teachers and their schools), endowing them with an open license has little meaning, given the risk that usability barriers will prevent (or at least severely inhibit) adoption.

In the case of the 4Ts game, there are at least four facets that need to be considered in order to make sure the game can be (re-)used by teachers in schools, teacher training institutions and the like: the format, the content, the software, and the hardware.

3.1   First Facet: Format

As mentioned above, the 4Ts game has been developed in three different formats. The paper format allows the teacher team playing the game to manipulate the cards and "see" the design as it is being produced on the board. In our preliminary experiments with this version of the game [22] it was soon clear that teachers welcomed the possibility to use the cards as mediating artefacts of their discourse around the design choices. The game stimulated collaboration among the teachers around the design, and this is per se an interesting result, given that designing for student learning is all too often an individual task. The collaboration, in turn, triggered reflection on the content of the cards, that is, the way Tasks, Teams, and Technology can be combined and laid on the board to form a coherent collaborative technique.

In terms of user-friendliness, this game format is the easiest to use because it does not require any complex technological set up and there is no interaction envisaged with the technology.

However, the need to have a tutor at hand to provide feedback and guidance on design choices poses a limit on the usability of the paper format. In addition, the persistence of the set up is generally limited to a single game session. In the above-mentioned projects where the game was tested, the game was used by teacher groups in different European countries and the constant presence of a teacher trainer could not be guaranteed. Hence the decision to implement the digital and hybrid versions capable of providing feedback, e.g. when a card is put in the wrong position of the board or to advise teachers when they get stuck and do not know how to proceed.

Both these formats have their affordances and limitations, in terms of usability. As mentioned above, when playing the game in its full-digital format, an Interactive Whiteboard (IWB) is highly desirable because playing the game on a PC would unlikely trigger the desired collaborative dynamics. IWBs are now found in many European schools, and teachers can often gain access to them outside class hours (this, at least, was the case in the two European projects where the game was tested[6]). In our experiments with the hybrid and digital versions of the game [16], the usability of the digital version was judged positively by teachers. However, the full-digital version does not offer the 'mediating artefact' power of physical cards and board.

By contrast, the hybrid version incorporates the chief technological affordances of the digital format, like the possibility to receive real-time feedback or suggestions, along with the advantages gained from manipulating the physical cards and positioning them on the paper board. Here, technological component is limited to a laptop positioned close to the board (Fig. 3). Although in our experiments the hybrid game was also evaluated positively for usability [16], it was regarded as slightly more cumbersome in comparison with the digital one. This was due to some difficulties caused by the setup, which sometimes turned out to be somewhat unsteady, e.g. due to accidental collisions with the camera or the table.

In summary, the availability of the game in three different formats allows teachers to choose the one best suited to the specific context/school, thus potentially widening adoption.

3.2 Second Facet: Content

As stated, 4Ts Game cards contain a text with prompts / indications on how the card (be it a Technique, Task, Team or Technology card) can be coherently combined with the other cards on the board. Teachers are expected to read these indications and make their choices accordingly. For example, if the learning task students are expected to carry out is 'debating', then the card is preferably combined with multi-person groups (a Team made up of one person cannot carry out this task) and videoconferencing systems (or face-to-face settings). Conversely, if the task is to 'study', individual work is deemed preferable (although studying in pairs or in a group is also possible) and the technology needed consists in the material to be studied (and some type of communication technology when the task is carried out in online groups). These 'rules of the game' are made explicit on each card but also incorporated in the digital component of the game as part of feedback (see the 'Third Facet: Software' section below.

As not all teachers are necessarily fluent in English, provision has been to facilitate translation of the card content (originally in English): card text is not embedded in the game code, but rather separately stored in a Google sheet. In the above-mentioned

---

[6] The schools involved were located in Belgium, Bulgaria, Cyprus, Greece, Italy and Spain.

projects, this feature was essential as in the different experiment contexts involved (particularly Bulgaria, Greece and Italy) the need clearly emerged to have the cards in end-user languages. The cards were translated by English-fluent teachers who volunteered their services. This task did not require any coding skills, but only respect for the positioning of the text in the spreadsheet cells, strict adherence to the original text, and consistent adoption of terminology. Thus, one of the frequently mentioned barriers to the adoption of OER, that is, the lack of resources in languages other than English [22], was addressed and overcome.

Moreover, it should be noted that the degree of flexibility provided by this simple expedient has much more to it than just facilitating production of translated game contents. Actually, it also provides the possibility for localization to counter cultural barriers: for example, aspects of content that are unsuitable in one culture can be modified instead of being literally translated, and some non-trivial repurposing can also be done. This surfaced in response to emergent needs within the two projects, where the remits of game use included focus on inclusive pedagogical approaches. Here, the content of the cards was more extensively modified by enriching the text with 'inclusion tips' concerning inclusive potential. The purpose of these tips is to explain how each Technique, Task, Team or Technology should be used in an inclusive manner. An example of a card with 'inclusion tips' is provided in Fig. 8. Thus, thanks to this feature, the content of the cards can be easily customized, according to the specific needs of the contexts. This demonstrates the potential for adoption in training contexts that address different issues.

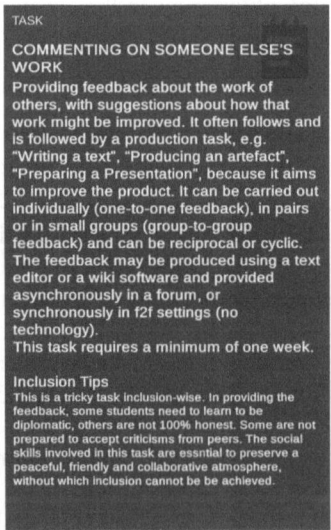

Fig. 8. Example of a Task card with related 'inclusion tips'

In addition, there is another form of open content embedded in the game: the so-called collaborative techniques. These are fully fledged "design patterns" [23, 24] that, once understood and appropriated by a teacher, can be reused in a number of different educational contexts to suit different learning aims and disciplinary contents. For example, teachers can use the well-known Peer Review technique to engage students in a three-phase collaborative pattern of activity. Here, two groups of students working in parallel each produce an artefact. Then, in the second phase, the groups swap artefacts to provide each other with feedback. Finally, in the third phase, each group, based on the feedback received, makes modifications to their original artefact with a view to improving it. This technique can be used effectively in any discipline, provided the teacher applies it appropriately. This practice, according to some authors, can be termed an Open Educational Practice [25].

3.3 Third Facet: Software

As far as open software is concerned, the open-source paradigm [26] is embraced world-wide and extensively adopted by several software developer communities. This paradigm is underpinned by principles that are similar to those sustaining open education.

In the case of the digital and hybrid 4Ts Game, openness is a distinguishing factor. The game architecture comprises three layers [17].

The top layer, i.e. the user interface, handles the board and the cards, and although it is implemented in Unity™ (a proprietary game engine), its code is open source. To implement the augmented features of the hybrid version, the OpenComputerVision library of ArUco markers was used[7]. These are printable square markers composed of a wide black border and an inner binary matrix that acts as identifier. Basically, these markers work like QR codes but are smaller and openly available.

The middle layer handles the business logic: system initialization, persistence management, syntax checks, output formatting, etc. This layer is implemented in C#, whereas queries and responses returned to the business logic are expressed in XML syntax.

The bottom layer implements the rules describing how the cards can be combined on the board and performs all the checks needed to identify errors in gameplay. This layer is the game knowledge base and it is implemented in SWI-Prolog. The code is open source and hosted on GitHub. The knowledge base is located in a separate network node (a cloud server) and can serve different interface clients in parallel. From an openness perspective (customizing the game), Prolog programming competence is needed to add cards or change the rules of the game.

The game runs on macOS or Windows with the latest OS and with at least 8GB of RAM. The complete user guide is provided in [27]. Moreover, Appendix 2 of [28] contains the complete technical documentation for developers, with all the information needed to customize the game software. All code is released under a General Public License.

As mentioned in the introduction, the open science requirement applying to all outputs from the two projects is in line with the principles of equity and democratization inspiring the whole open education movement. However, few teachers possess the skills to modify or customize a complex software system like the 4Ts Game. Hence, personalizable features have been built into the system, at least for those aspects that our

---

[7] The "OpenComputerVision" library of ArUco markers by Oleg Kalachev can be found here: https://github.com/okalachev/arucogen.

experiments revealed as potentially requiring personalization or localization. So, similarly to what has been done with the card content (see section above), some blank wild cards were included in the game so that users can add whatever Task, Team, Technology or Technique they wish. Hence, teachers who do not possess programming skills can significantly customize the game to meet their needs.

3.4 Fourth Facet: Hardware

As far as hardware is concerned, once again the game developers made an special effort to meet the needs typically arising in the target area, i.e. primary or secondary schools. Here, expensive equipment and facilities are often beyond reach.

For the paper-based game obviously no hardware is required; only the board and cards need to be printed. These are freely available online in PDF format. Hence, they can be easily downloaded from the game website[8], printed out on the school printer and cut out with scissors. The board can also be printed 'in house' using A4 paper, but in this case the sheets will need to be assembled to form the whole board. Alternatively, a more durable long-lasting board can be printed out at relatively low cost by a professional printing service; cardboard, fabric or low-density polyethylene cardboard can all be adopted for this purpose.

As already mentioned, the digital version of the game is expected to run on an IWB or a smart TV with a touch screen. The hybrid format requires printing out of the cards and board, just as for the paper version. These are laid out on a table, to which a camera on an extender stick is affixed. The camera image of the table from above does not need to be high definition, so an inexpensive model is sufficient. A cheap microphone holder (available online for less than €30) is ideal for use as an extender stick.

Thus, while in principle all three versions of the game can be adopted even in schools with limited resources, each school is free to choose the set up which fits better with its own aims, equipment and facilities.

4 Conclusive Remarks

A plethora of studies has focused on OERs and their definition, as well as on enablers and barriers to their adoption in diverse educational contexts. In parallel, research agencies funding projects at national and international level have adopted policies intended to foster Open Education practices [29]. While there is no disagreement that adoption needs to be encouraged [30], "open education often does not live up to its own vision: in practice, unequal access to communications technology, unequal distribution of basic study skills, and unavailability of resources in certain languages mean that open approaches can act as a force for exclusion rather than inclusion" [31].

In this contribution, we discussed the approach adopted in developing a serious game for training school teachers that has been developed in two Erasmus + projects. The belief behind this approach is that when developing advanced technological tools for use in schools, open-source software and open intellectual property licenses are not

---

[8] https://sites.itd.cnr.it/4TsGame/.

enough. Uptake should be facilitated with a very pragmatic approach by developing material which has built-in features for localization and adaptation to different contexts, and do not pose unrealistic equipment or facility requirements. It is well known that not all schools have plentiful technological and economic resources, and face a range of educational challenges. For this reason, both the digital and the hybrid versions of the game described in this paper were developed with an eye to low-floor requirements and hardware that schools are likely to have or can purchase with limited budget. For example, the hybrid version could easily have been implemented for digital tabletops or touch tables, but this was deliberately avoided because these technologies are usually not part of school equipment. As we have mentioned, the resulting game set up presents some weakness, so we believe further research needs to be conducted to develop and deliver innovative educational solutions that leverage technology which can be realistically adopted in schools on a large scale.

Finally, it should be recognized that ICT competences are frequently not part of teachers' skill sets. Hence, even before making sure that software can be modified and personalized, it is important that proposed tools are flexible in terms of format, content, software and hardware so that adoption and adaptation require little effort and regular teacher competences.

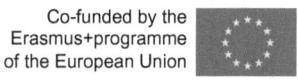

Co-funded by the
Erasmus+programme
of the European Union

Acknowledgements. The work presented in this paper has been carried out with the support of the PLEIADE and SuperRED projects, co-funded by the Erasmus+ Programme of the European Union (respectively agreement numbers 2020–1-IT02-KA201–080089 and 2021–1-IT02-KA220-SCH-000034442). The European Commission's support for the production of this publication does not constitute an endorsement of the contents, which reflect the views only of the authors, and the Commission cannot be held responsible for any use which may be made of the information contained therein. The funders had no role in the design of the study, in the collection, analyses, or interpretation of data, in the writing of the manuscript, or in the decision to publish the results.

References

1. Olimpo, G., Persico, D., Sarti, L., Tavella, M.: On the concept of database of multimedia learning material. In: McDougall, A., Dowling, C. (eds), Proceedings of. Fifth World Conference on Computers in Education (WCCE 90), pp. 431–436. Elsevier, Amsterdam (1990)
2. Wiley, D., Bliss, T.J., McEwen, M.: Open educational resources: a review of the literature. Handbook of research on educational communications and technology, pp.781–789 (2014)
3. Hilton, J., Wiley, D., Stein, J., Johnson, A.: The four 'R's of openness and ALMS analysis: frameworks for open educational resources. Open Learn. **25**(1), 37–44 (2010)
4. Hilton, J.: Open educational resources and college textbook choices: a review of research on efficacy and perceptions. Educ. Tech. Res. Dev. **64**, 573–590 (2016)
5. Tlili, A., Garzón, J., Salha, S., et al.: Are open educational resources (OER) and practices (OEP) effective in improving learning achievement? A meta-analysis and research synthesis. Int. J. Educ. Technol. High. Educ. **20**(1), 54 (2023). https://doi.org/10.1186/s41239-023-00424-3

6. Feldstein, A., Martin, M., Hudson, A., Warren, K., Hilton III, J., Wiley, D.: Open textbooks and increased student access and outcomes. Eur. J. Open, Distance E-Learn. **15**(2) (2012)

7. Mullens, A.M., Hoffman, B.: The affordability solution: a systematic review of open educational resources. Educ. Psychol. Rev. **35**(3), 72 (2023)

8. Atenas, J., Havemann, L.: Questions of quality in repositories of open educational resources: a literature review. Res. Learn. Technol. **22**, 20889 (2014)

9. Tlili, A., Nascimbeni, F., Burgos, D., Zhang, X., Huang, R., Chang, T.W.: The evolution of sustainability models for Open Educational Resources: Insights from the literature and experts. Interact. Learn. Environ. **31**(3), 1421–1436 (2023)

10. Beaven, T.: Dark reuse': An empirical study of teachers' OER Engagement. Open Praxis **10**(4), 377–391 (2018). https://doi.org/10.5944/openpraxis.10.4.889

11. Kortemeyer, G.: Ten years later: why open educational resources have not noticeably affected higher education, and why we should care. Educ. Rev. 48(2) (2013)

12. Baas, M., Admiraal, W., van den Berg, E.: Teachers' adoption of open educational resources in higher education. J. Interact. Media Educ. **2019**(1), Article 9 (2019)

13. Cox, G.J., Trotter, H.: An OER framework, heuristic and lens: tools for understanding lecturers' adoption of OER. Open Praxis **9**(2), 151–171 (2017). https://doi.org/10.5944/openpr axis.9.2.571

14. UNESCO: Recommendation on Open Educational Resources (OER). The General Conference of the United Nations Educational, Scientific and Cultural Organization (UNESCO), Meeting in Paris from 12 to 27 November (2019). https://unesdoc.unesco.org/ark:/48223/pf0 000373755/PDF/373755eng.pdf.multi.page

15. Mishra, S.: Open educational resources: removing barriers from within. Distance Educ. **38**(3), 369–380 (2017)

16. Pozzi, F., Ceregini, A., Ivanov, S., Passarelli, M., Persico, D., Volta, E.: Digital vs. hybrid: comparing two versions of a board game for teacher training. Educ. Sci. **14**, 318 (2024). https://www.mdpi.com/2227-7102/14/3/318

17. Ceregini, A., Persico, D., Pozzi, F., Sarti, L.: The 4Ts game to develop teachers' competences for the design of collaborative learning. In: Burgos, D., Cimitile, M., Ducange, P., Pecori, R., Picerno, P., Raviolo, P., Stracke, C.M. (eds.), Higher Education Learning Methodologies and Technologies Online, First International Workshop, HELMeTO 2019, Revised Selected Papers. Communications in Computer and Information Science (CCIS), vol. 1091, pp. 192–208. Springer, Springer, Heidelberg. https://doi.org/10.1007/978-3-030-31284-8_15

18. Pozzi, F., Ceregini, A., Persico, D.: Designing networked learning with 4Ts. In: Cranmer, S., Dohn, N.B., de Laat, M., Ryberg, T., Sime, J.A. (eds.) Proceedings of the 10th International Conference on Networked Learning 2016, pp.210–217 (2016). http://www.networkedlearni ngconference.org.uk/abstracts/pdf/P15.pdf

19. Pozzi, F., Persico, D.: Sustaining learning design and pedagogical planning in CSCL. Res. Learn. Technol. **21** (2013)

20. Pozzi, F., Persico, D. (eds.): Techniques for Fostering Collaboration in Online Learning Communities. Theoretical and practical perspectives. IGI Global-Information Science Reference, Hershey, PA (2011). https://doi.org/10.4018/978-1-61692-898-8

21. Persico, D., Dagnino, F. M., Manganello, F., Passarelli, M., Pozzi, F., Nikolova, N., Lonigro, M.: SUPPORTING TEACHERS'PROFESSIONAL DEVELOPMENT ON INCLUSIVE LEARNING DESIGN: A CASE STUDY OF AN ERASMUS+ PROJECT. In INTED2023 Proceedings, pp. 7456–7464. IATED. (2023).

22. Pozzi, F., Manganello, F., Persico, D.: Collaborative learning: A design challenge for teachers. Education Sciences. **13**(4), 331 (2023)https://doi.org/10.3390/educsci13040331

23. Baggetun, R., Rusman, E. Poggi, C.: Design patterns for collaborative learning: from practice to theory and back. In: Cantoni, L., McLoughlin, C. (eds.) Proceedings of ED-MEDIA

2004--World Conference on Educational Multimedia, Hypermedia & Telecommunications, pp. 2493–2498, Association for the Advancement of Computing in Education (AACE) Lugano, Switzerland (2004). https://www.learntechlib.org/primary/p/12374/. Accessed 21 Mar 2024

24. Goodyear, P., Retalis, S. (eds.): Technology-Enhanced Learning: Design Patterns and Pattern Languages. Sense Publishers, Rotterdam, The Netherlands (2010)
25. Cronin, C., MacLaren, I.: Conceptualising OEP: a review of theoretical and empirical literature in Open Educational Practices. Open praxis **10**(2), 127–143 (2018)
26. Wu, M.W., Lin, Y.D.: Open Source software development: an overview. Computer **34**(6), 33–38 (2001)
27. Bicocchi, M., Ceregini, A., Innocenti C., Persico, D., Polsinelli, P., Pozzi, F., Sarti, L.: The Hybrid I4Ts Game (PLEIADE Intellectual Output No. 2) (Revised version) (2022). https://doi.org/10.17471/54014
28. Passarelli M., Pozzi, F., Persico, D., Volta, E.: Impact amplification kit (PLEIADE Intellectual Output No. 6) (2023). https://doi.org/10.17471/54021
29. Cronin, C.: Openness and praxis: Exploring the use of open educational practices in higher education. Int. Rev. Res. Open Distrib. Learn. **18**(5) (2017)
30. Otto, D.: Adoption and diffusion of open educational resources (OER) in education: a meta-analysis of 25 OER-projects. Int. Rev. Res. Open Distrib. Learn. **20**(5), 122–140 (2019)
31. Farrow, R.: A framework for the ethics of open education. Open Praxis, 8(2), 93–109 (2016). https://www.learntechlib.org/p/173546/. Accessed 28 Mar 2024. (International Council for Open and Distance Education, 2016)

Author Index

A. Aldini (Ed.): SEFM 2023, LNCS 14568, p. 169, 2024.
https://doi.org/10.1007/978-3-031-66021-4